FAO FISHING MANUALS
TUNA FISHING WITH POLE AND LINE

Tuna Fishing with Pole and Line

edited by

M Ben-Yami,
Fishery Industries Division, FAO

Published by arrangement with the
Food and Agriculture Organization of the United Nations
by **Fishing News Books Ltd**
1 Long Garden Walk, Farnham
Surrey, England

© FAO 1980

The copyright in this book is vested in the Food and Agriculture Organization of the United Nations for which Fishing News Books Ltd. acts as publisher. The book may not be reproduced, in whole or in part, by any method or process, without written permission from the copyright holder. This applies in particular to photo copying of the designs and plans. Applications for any desired permission should be addressed to the Director, Publications Division, Food and Agriculture Organization FAO, Via delle Terme di Caracalla, Rome, Italy, accompanied by a detailed explanation of the purpose and extent of the reproduction desired.

British Library C I P Data

Tuna fishing with pole and line.—(Food and
   Agriculture Organization. Fishing manuals).
   1. Tuna fisheries
   2. Pole and line fishing
   I. Series
   639'.27'58   SH351.T8

ISBN 0 85238 111 5

Typeset by
Information Services Division,
Brown Knight & Truscott Ltd.,
Tonbridge and London

Printed in Great Britain by
Whitstable Litho Ltd., Whitstable, Kent

CONTENTS

LIST OF FIGURES

PREFACE

This manual has been prepared at the Fisheries Technology Service, FAO Fishery Industries Division on the basis of material supplied from many sources, as the bibliographic acknowledgements listed indicate. The main source has been a manuscript from the Japan Tuna Fisheries Cooperative Federation written by Teruo Konagaya in which he dealt with pole-and-line fishing for tuna, mainly in relation to Japanese experience. FAO is grateful to him for his valuable contribution and to W Reed for his description of the Polynesian variant of pole-and-line fishing, as well as to M Mukundan for his contribution on the pole-and-line fishing carried out by the fishermen of Lakshadweep, and K Sivasubramaniam for his description of this fishery in Sri Lanka. Other valuable contributions came from Captain G Pajot. Dr W Fischer who checked the fish names, and Ms A Barcali and Mr S Maugeri drew many of the line drawings.

Much of the material in this manual has been derived from the experience and reports of FAO personnel working on tuna fishing development in various parts of the world; also from personal communications and from many articles and other published material, which can be found listed in the bibliography.

This manual attempts to provide information and advice to those fishermen, fishing technologists, fishing instructors and extension workers for whom such pole-and-line fishing is new, as well as for those who wish to improve their present technique.

It is aimed at rather small-scale operations and is not intended to 'carry coals to Newcastle', which is why the large-scale Japanese and American fisheries are dealt with in a somewhat cursory manner.

CHAPTER 1

INTRODUCTION

1.1 Historical note

Man has fished for tuna since the dawn of history, as indicated by the fact that bones of skipjack have been found in Stone Age mounds. The remains of barbless hooks made from horn and of age-old dugout canoes further point to hook-and-line fishing in ancient times. Japanese tradition and records, for example, provide a continuing story of the development of tuna fishing, including the introduction of metal hooks in Japan by the 8th century AD, and the use of nets in skipjack fishing in the 12th century. The Japanese records also indicate that angling became the main tuna fishing method and that the use of live bait, artificial lures and water spraying with bamboo dippers took place during the period from 1600 to the latter part of the 19th century.

The chumming of tuna with live bait and fishing with pole-and-line have been used since remote times by fishermen of Madeira and the Azores. Portuguese fishermen from the Azores introduced this fishing method into California, USA, where it developed into an important industry.

At first, dugout canoes, row boats and sailing boats were used, and large sailing vessels were introduced during the 19th century. In the early years of the present century the introduction of mechanical power to the fishing fleets led also to the modernization of the pole-and-line vessels and thus to the expansion of the fishery over wider and wider areas.

1.2 New fishing grounds for small-scale pole-and-line fishery

While many tuna fisheries are well established over well-known inshore, offshore and oceanic fishing grounds, there are good reasons to believe that other grounds exist. There is need, therefore, to survey prospective areas. A thorough survey involves the investigation of oceanic conditions; the fish stocks and their distribution; the movement patterns of the schools, including feeding and spawning migrations, and the seasonal changes in the area in relation to their influence on the fish. Such a survey requires research vessels, facilities and oceanological expertise and that of marine biologists, but these are not always available. However, it should be borne in mind that adequate starting information can be obtained for pole-and-line fishing without complex investigations, especially in inshore and offshore areas not too distant from national bases. Such information can be collected from local fishermen, who are generally very observant of natural phenomena associated with fishing and fish, and augmented and verified through practical, exploratory fishing trips in the areas concerned. An essential part of such an investigation must be the location of suitable stocks of live bait, a prerequisite for any new pole-and-line fishing venture.

1.3   Live bait

It will be noticed that considerable attention is paid in this manual to live bait. This is because fishing for tuna with pole-and-line depends completely on a constant supply of live bait and on its survival until needed for the fishing operation. The manual therefore deals at some length with the techniques used for catching and preserving live bait and with the various species caught in different parts of the world, including their specific habitats. Relative to this is the fact that much live-bait fishing is conducted with the aid of artificial light. The many methods used in light fishing have been dealt with in detail in the FAO Manual *Fishing with Light*. The reader should obtain a copy of this book if he wishes to study in more detail the light-fishing methods referred to in the present manual with respect to live-bait fishing. (See Bibliography).

1.4   Scope for development

Whatever type of vessel is used, be it a dugout canoe, sailing boat, small or large motorized ocean-going vessel, pole-and-line fishing for tuna remains basically the same—to locate, then attract by chumming and spraying water and finally to catch these fish using poles and hooked lines. The opportunities for successfully taking up such fishing on a small or medium-scale are widespread. This is because of the world-wide distribution of the various species of tuna, some of which are found mainly in inshore waters while others are found in offshore and distant ocean waters.

The catches of the various classes of pole-and-line fishing boats reflect this. The Japanese, for example, have found that their large vessels which fish the more distant waters, catch mostly skipjack, with a smaller proportion of albacore (about 20 percent) and only some 4 percent of small species. Their medium size boats with a medium range of operation catch still more skipjack, and only about 10 percent of albacore, and 9 percent of small species. The small inshore boats catch mostly frigate mackerel, then skipjack (up to about 40 percent), some 4 percent of small species and only 2 percent of albacore. Of course, these figures vary for fishing boats of other countries, depending on the predominance of the species in the waters concerned. However, skipjack is one of the most widespread species found in all tropical and subtropical waters so that it is likely to form a substantial part of the catch of any fishing boat operating in such areas.

The current extension by coastal countries of their jurisdiction over the fisheries to 200 miles from shore is changing the fisheries situation because many of the distant water fishing grounds will come within the new national limits and thus become inaccessible to the big foreign tuna fishing vessels, other than on a limited, negotiated basis. This may well lead to situations where the employment of smaller and much less expensive boats in the pole-and-line fishery by fishermen of the coastal countries would be a feasible alternative for the exploitation of their national resources.

As already pointed out, there are many other prospects for pole-and-line tuna fishing, since these fish abound in the inshore and offshore waters of a number of developing countries. The exploitation of these resources is often limited by the lack of know-how on the part of local fishermen and by lack of investment funds. In other countries—the Maldives or Cape Verde, for example—the pole-and-line fishery is well established and makes an important contribution to the country's economy. However, even in such cases, improvement may be brought about with the help of investment capital and the introduction of a higher level of technology. One of the purposes of this manual is to point the way to such improvements.

Obviously, a manual of this kind cannot present a ready-made choice of boat, gear, equipment, *etc* or of fishing area. The choice of these must be made by the fishermen and fishing technologists concerned, according to their knowledge of local conditions and the available fishing grounds, and by their level of technology and ability to finance new or extended operations. The objective of this manual, therefore, is to survey the main range of vessels, fishing gear, and auxiliary equipment, and the operations involved in pole-and-line fishing, so as to enable small- and medium-scale fishermen in the developing countries to see how they could enter such a fishery. This can be an important step towards increasing their earnings and improving their standard of living. We hope that these and the following pages will be of practical assistance to such people.

CHAPTER 2

MAIN SPECIES FISHED

The pole-and-line fishermens' main catch is skipjack (*Katsuwonus pelamis*), albacore (*Thunnus alalunga*), small tunas, such as frigate mackerel (*Auxis* spp), and dolphin fish (*Coryphaena* spp); also yellowfin (*Thunnus albacares*), young fish of other species of large tuna, bonitos (*Sarda* spp) and little tuna (*Euthynnus* spp). All these are widely distributed in the seas and oceans of the world.

2.1 Skipjack

2.1.1 BIOLOGY AND DISTRIBUTION

As mentioned earlier, the skipjack comes first in pole-and-line fishing. It is a fast-swimming torpedo-shaped fish, chiefly inhabiting tropical and subtropical waters where the temperature ranges from 17°C to 30°C. The fish starts spawning when it is one year old, beginning with about 100,000 eggs each spawning period and increasing as it grows older to some 2 million eggs per spawning period.

The fish spawns several times in each spawning period, the eggs hatching out about four days after fertilization. The larvae are found over wide areas in the Indian, Pacific and Atlantic Oceans. The concentrations of the fry in the Pacific increase from east to west, being especially heavy in the area between 10°N and 10°S. As the fish grow they spread out, and in the Pacific they have migrated by the age of two from the areas where they were spawned, into the Eastern Pacific. Skipjack of four years or more return to equatorial waters to spawn.

Oceanographic conditions have considerable influence on skipjack. According to the experience of Japanese fishermen and researchers, the fish gather in commercial concentrations in latitudes 40°N to 40°S, especially where the colour of the sea is blue to blue-green. The transparency of the water as measured with a Secchi disc exceeds 20m and the temperature in its upper layer is from 17°C to 30°C.

Schools are particularly dense in the equatorial zones of the central and western Pacific. At present, pole-and-line fisheries are mainly active in the western Pacific (from Japan to the tropical area); in the offshore waters of the eastern Pacific (from California to Chile); in the waters around the Hawaiian Islands; in the Indian Ocean (mainly Lakshadweep, Maldives and Sri Lanka) and in the Atlantic Ocean (off the coast of West Africa, including the regions of Cape Verde, Azores, Madeira and Canary Islands, off the east coast of the USA and in the Caribbean). There still seems to be much scope for exploiting the fish in the rest of the area of their distribution.

2.1.2 FOOD AND BEHAVIOUR

The fish feed on a wide range of small marine animals, varying according to the prevalence of the prey species. When feeding they surround their prey, a behavioural feature which, by preventing the live bait from dispersing, is helpful in pole-and-line fishing. While skipjacks are attracted to live bait generally they do show preferences for some species, again according to the prevalence of food available in the area.

Skipjack swim up currents but down wind-driven waves. Where schools are attached to underwater shoals and banks, fish of different sizes tend to swim at different depths. Migrating schools consist of fish of uniform size.

2.2 Albacore

Albacore are found in temperate and tropical waters between 45°N and 45°S. In the Pacific they abound in the area between 130°E and 180° longitude and 20°N to 40°N latitude; also off the Hawaiian Islands and the West Coast of the USA. It is generally agreed that the stock in the northern hemisphere is different from that in the southern and that each has its own spawning grounds and area of distribution. That in the north is distributed mainly between 10° and 40°N, particularly in the waters between the polar front and the Equatorial Counter Current. Fish of less than 100cm body length occur in the waters north of the subtropical convergence zone while larger fish migrate to the areas of the North Equatorial Current and the Equatorial Counter Current.

The albacore in the southern hemisphere are distributed between the equator and 40°S but fishing for them takes place mainly in the area between 10° and 30°S latitude and 150°E and 120°W longitude. They are caught by pole-and-line and by trolling off Chile and around North Island, New Zealand. The larger fish (over 90cm) dominate in waters north of 30°S while south of that latitude the fish are mostly smaller.

In the Indian Ocean the albacore are mainly found south of 10°S latitude, the larger fish being in the areas north of 30°S, and the small, immature fish south of that latitude. In general, the albacore abound at both sides of 30°S latitude and off the tip of Africa in waters adjacent to the Atlantic.

In the Atlantic they are found between 45°N and 40°S, a wide area where pole-and-line fishing has long been conducted. Immature albacore and bluefin tuna have been fished in the Bay of Biscay since 1945. The fishing was extended to the waters off Senegal around 1955 where, however, the bulk of the catch consists of yellowfin tuna and skipjack. In subsequent years, fishing by European vessels has extended down to the Gulf of Guinea, with purse seines gradually taking the place of pole-and-line vessels.

2.3   Smaller species

2.3.1   FRIGATE MACKERELS

Frigate mackerel (*Auxis thazard*) and bullet mackerel (*Auxis rochei*) are
found in the waters off Hokkaido to the Philippines, frigate mackerel also
being found around the Hawaii area, along the west coast of North
America, in the Mediterranean and in the Atlantic. It is caught throughout
the year in the Southern Pacific and from early summer to late autumn in
the Northern Pacific. Good catches of frigate mackerel usually indicate an
opportunity to make good catches of other warm water species. Frigate
mackerel prefer clear water where the currents are strong so that they tend
to gather in the vicinity of capes.

2.3.2   BONITOS

The oriental bonito (*Sarda orientalis*) is distributed in the temperate waters
of the Atlantic, Indian and Pacific Oceans. The Pacific bonito (*Sarda
chilensis*) is fished off California and Chile while the Atlantic bonito (*Sarda
sarda*) is caught in the Atlantic Ocean and the Mediterranean Sea.

2.3.3   LITTLE TUNAS

Little tuna (*Euthynnus* spp) are warm water fish, mainly found in the
inshore areas. The eastern variety (*Euthynnus affinis*) are fished in the area
south of Formosa and Honshu, around India and in the Indian Ocean, off
Australia and in the Red Sea as well as around Hawaii and the southwest
Pacific Islands.

CHAPTER 3

THE VESSELS

While the same principles of tuna pole-and-line fishing apply throughout the world, there is a wide range of vessels that have been designed or converted for the job. It is not possible, of course, to examine all these in this manual so a few of the main types have been selected, with emphasis on the smaller craft more appropriate for developing fisheries.

The size of the vessels and the amount of live bait they have to carry are dictated by the distance at which these vessels operate. The size in turn determines the engine power, the capacity of the refrigeration equipment and the range of radio and navigation aids, *etc*. Almost all distant water fishing vessels have radar, radio direction finders, long-range radio navigation systems, echo sounders, ship-to-shore and ship-to-ship radio, gyrocompasses, track recorders *etc*, as standard equipment.

In principle, there are two distinctly different types of distant water pole-and-line tuna fishing vessels. One is the *tuna clipper*, which prevails in the Eastern Pacific Ocean and the North and Central Atlantic Ocean, and the other is a type developed in Japan which prevails in the Far East.

3.1 Tuna clippers

3.1.1 GENERAL FEATURES

Tuna clippers are a class of specialized vessel which was developed on the West Coast of the USA but now equally employed in the Atlantic, mainly by French and Portuguese fishermen. Their numbers and importance decreased considerably following the development of the tuna purse seining fishery during the 1950s and 1960s. The recent introduction of automated tuna angling machines may revitalize the US pole-and-line fishery. The distinct features of the clippers involve superstructure forward of midships, deck-situated live-bait tanks aft and outboard fishing racks. The angling is done from around the stern (*Fig 1*).

Specifications of tuna clippers show that their overall length ranges from about 20m to 45m, the beam from 6m to 10m and the draft from 2.5m to 4.7m. The vessels may be from about 60 to 300t nett and be made from steel or wood. They have diesel engines of 200 to 1,200hp, capable of a cruising speed of 9 to 12 knots. With some exceptions, they are fitted with brine tanks and mechanical refrigeration, and as the specifications indicate, the bulk of the clippers are built for fishing in distant waters (*Fig 2*). The

Fig 1   Stern arrangement of a World War 2 US naval tug, converted to exploratory tuna clipper

trips in fact, may last from one to three months. The crew, depending on the size of boat, type of fishing, and the extent of mechanization of equipment on board, may number from 9 to 21 men or even more.

3.1.2   FISHING RACKS

One important and distinct feature of the tuna clippers is the fishing rack, *ie*, removable outboard platforms on which the fishermen stand during fishing operations.

General practice is to fit a rack around the stern of the boat. Some vessels have racks across the stern and to supplement this, one or two one-man racks which hang over the rail and are supported by the side of the vessel. These may be used on either side of the boat at the most convenient point.

Racks are usually constructed of galvanized iron pipe with a rail above knee height and another about 75mm above the floor. The latter acts as a useful toe hold. The floor is of bars, wooden slats, or heavy gauge expanded metal. The width of the rack should not be less than 85cm.

An important consideration when fitting a rack is the height above the floor of the rack of the rail, over which the fish must be landed. If this is too great it is extremely difficult to bring the tuna aboard. The most suitable height is between 80 and 90cm. Also important is the height of the floor of the rack above the water. This will vary with the loading of the vessel, but the lower the floor in relation to the water, the easier it is to pole tuna.

The stern rack may be attached by hinges at the inboard edge of the floor and held in position by short lengths of chain. It can then be pulled up, with the floor flat against the stern, when not in use.

3.1.3 LIVE-BAIT WELLS

Another distinct characteristic of the tuna clippers is that the live-bait wells are installed on the after deck, mostly at deck level, and within easy reach of the chumming men.

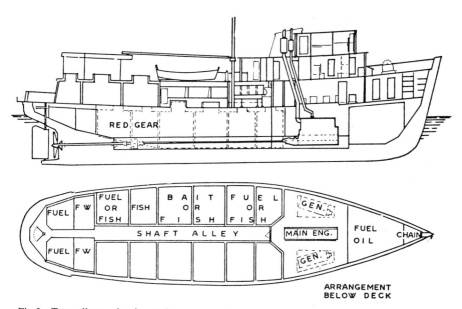

Fig 2   Tuna clipper, showing tank arrangement

3.2   Distant-water pole-and-line vessels of Japanese type

3.2.1 DISTINCTIVE FEATURES

These specialized vessels are generally of more than 200t light displacement (over 200grt) and over 35m in length, and are robust in construction, relatively fast and seaworthy. They are powered by diesel engines of 1,000 to 3,000hp. Their distinguishing features include a clipperbow, a protruding fishing platform, the so-called 'sponson', running on the bow and along the vessel's sides, deckhouse aft of the midships and 10 to 12 live-bait tanks or wells, each holding from 240 to 600kg of bait. The tanks are mounted under the deck to be secure in bad weather and to increase the vessel's stability.

Fig 3   Large Japanese type pole-and-line fishing vessel

They are used as fishholds on the return voyage. (*Figs 3* and *4*)

As in the case of tuna clippers, these vessels are equipped with freezing and refrigerating machinery, water sprays, and other special equipment for fishing, notably the recently developed automatic angling machines. The level of technology is similar to that of the American clippers, perhaps even more elaborate, although the crew of such a vessel may number as much as 45. This includes the skipper, chief fisherman, mate, chief engineer, engineer and wireless operator, all of whom take part in fishing as required, with the rest of the crew.

3.2.2   THE SPONSON

The 50–60cm wide sponson, unlike the fishing racks of the clippers, is an integral part of the vessel's hull. It may be located on either port or starboard, depending on local custom (*Fig 109*) and on bigger vessels, the platform is placed as low as possible to make it easier for the fishermen to flip the fish on board. Care has to be taken not to place the platform so low that it interferes with the vessel's sea-going qualities, however. The normal height, therefore, has been established at 1.0 to 1.5m above the water line. The boat in *Fig 10*, has, in fact, been criticized as having its bow sponson too high.

3.2.3   REFRIGERATION

Freezing and refrigerating equipment consists of two to four brine-freezer tanks with a capacity ranging from 20 to 60t a day at −18°C. They are installed near the live-bait tanks.

A semi-airblast 'sharp' freezer is installed in each vessel and has a capacity of 2 to 5t daily. The freezing temperature is −45° to −60°C.

When vessels are in warm water areas, the refrigerating equipment may be used to cool the water circulating in the live-bait tanks.

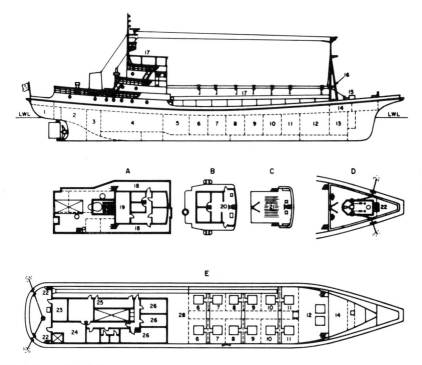

Fig 4   Large (314t) Japanese type tuna pole-and-line fishing vessel
A. Bridge deck;   B. Navigation bridge deck;   C. Compass deck;   D. Forecastle deck;   E. Upper deck
1.Aft store; 2.Fuel oil tanks (port and, starboard); 3.Fresh-water tanks (port and starboard); 4.Engine room; 5.Brine-cooler room; 6–11.Bait and brine holds; 12.Fish hold; 13.Freezing room; 14.Bosun's store; 15.Windlass; 16.Mast; 17.Awning; 18.Wooden plank; 19.Radio; 20.Wheelhouse; 21.Grating; 22.Bitt; 23.Steering engine; 24.Galley; 25.Battery room; 26.Crew space; 27.Line belt conveyor; 28.Planked deck

3.2.4   THE SPRAY SYSTEM

A steel pipe of 10mm diameter is mounted above the fishing platform with a brass spray nozzle fitted every 70cm (*Fig 5–6*). These sprinklers can be adjusted so that the spray falls where required. An adaptation of this spray device that can also be used in small or medium sized boats is shown in *Fig 7*. It consists of a short section of rubber hose wired to a water point on the pipe and a short piece of bamboo fitted into the end of the hose, the knot joint of the bamboo being at the outside end. This knot is then pierced as required to permit the water to spray through. If an all-metal nozzle is used, the spraying effect is achieved by flattening the end of the nozzle pipe.

Fig 5   Bow spray system on board a large Japanese pole-and-line vessel. Note individual sprinklers installed in special holes in the ship's plating

Fig 6   Stern spray system on the vessel in *Fig 5*. Note chum tank on port quarter

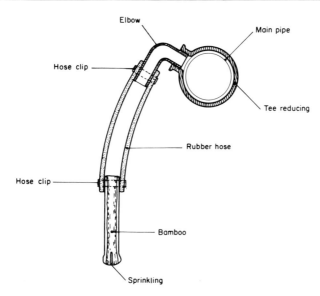

Fig 7   An easily made sprinkler

3.2.5   AUTOMATIC ANGLING MACHINES

Automatic fishing machines (*Fig 8*) are placed on the fishing side of the vessel, the number being determined by the length of the boat and the number of the crew. A special advantage is that the machines can be placed high up on the vessel's side. Each machine is remote controlled and has an

Fig 8   An automatic angling machine. The arrow indicates point of attachment for the pole

arm to hold the fishing pole and moves it up and down, agitating the artificial line. When a fish strikes, the machine instantly stops the movement and hauls the fish on board (*Fig 9*). The fish drops off and the machine throws the hook and line with its artificial lure back into the sea and starts the up and down jigging movement again.

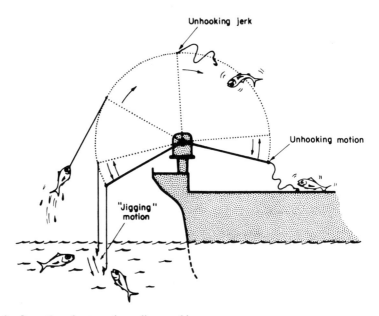

Fig 9   Operation of automatic angling machine

3.2.6   OTHER EQUIPMENT

A pumping system to carry live bait from the main to the chum tanks on the fishing deck involves vacuum suction and transfer by low air pressure. The suction hose is placed in the main tanks and sucks up the fish with sea water, carrying them to the small chum tanks, situated aft, amidships, and on the foredeck. There, the bait is kept during the fishing operation and within easy reach of the 'chummers'.

Standard equipment for these modern vessels includes thermometers for measuring surface and midwater temperatures, as well as the previously mentioned electronic equipment (see above), to search for fish and to receive information on oceanographic, metereological and fishing conditions, and to report to shore and to other vessels.

3.3   Medium-sized vessels of the Japanese type

Medium-sized vessels of 30 to 100grt, with 400 to 800hp diesels and a speed of up to 11 knots, are used in the Far East for the offshore fishery (*Figs 10,*

11 and *12*). They fish in waters 200 to 1,000 miles from shore, the cruises lasting from seven to 10 days, of which three or four are spent fishing. The catch amounts to five to 10t per cruise.

3.3.1 CONSTRUCTION, CREW AND EQUIPMENT

These boats, like the bigger ones, have clipper bows and sponsons. Their hulls are of wood or of fibre-reinforced plastic (FRP or GRP). They have five or six live-bait tanks installed amidships. The boats must have refrigerated fish-holds, and preferably, freezers. The crew may number 20 to 25. Equipment includes radar and Loran, thermometers, radio and radio telephone but so far, not the automatic angling machines.

Fig 10   A medium (57t) Japanese pole-and-line vessel; it has been criticised as having its bow sponson too high for easy angling

Fig 11   Japanese medium type pole-and-line vessel

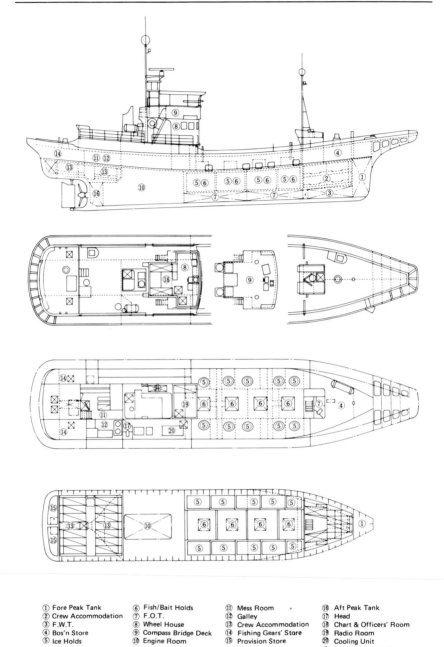

① Fore Peak Tank	⑥ Fish/Bait Holds	⑪ Mess Room	⑯ Aft Peak Tank
② Crew Accommodation	⑦ F.O.T.	⑫ Galley	⑰ Head
③ F.W.T.	⑧ Wheel House	⑬ Crew Accommodation	⑱ Chart & Officers' Room
④ Bos'n Store	⑨ Compass Bridge Deck	⑭ Fishing Gears' Store	⑲ Radio Room
⑤ Ice Holds	⑩ Engine Room	⑮ Provision Store	⑳ Cooling Unit
			㉑ Main Switchboard

Fig 12   Plan of Japanese medium type pole-and-line vessel

At the upper end of the 'medium-size' range are vessels such as those illustrated and described in *Figs 11 and 12*. At the lower end of this range we find a 39grt 19.5m overall length vessel, 4.4m beam and 2.05m draught. It has a cruising speed of 8 knots and a cruising range of 3,200km or about 10 days at sea, a 250hp main engine, and a 7hp auxiliary engine to drive the 100v AC generator of 5kW. It is equipped with radar, radio direction-finder, depth recorder and a two-way radio. The vessel is fully decked and fitted with a wheelhouse and flying bridge amidship. The well deck is forward of the bridge and has 12 hatches: three bait wells, an ice hold and four additional holds on each side. Two of the latter have 4t capacity water tanks, the other six being fish holds for carrying 12 tons of iced fish.

The bait wells have a natural circulation of sea water through holes 10.5cm (4.1in) diameter cut in their bottom planking. One well has wooden plugs, the other two have air-actuated metal shut-off valves, for closing the holes when the wells have to be made watertight. When they are flooded the bait fish are prevented from escaping by a bamboo screen fitted over the holes. The draught of the vessel governs the level of water in the wells, the circulation being dependent on the vessel's motion, but when it is at anchor or otherwise still, pumps are used to circulate the water. The wells are painted white, and are illuminated by a 40w lamp. When a large catch of tuna is made, wells are pumped dry and the catch is stored in them.

The vessel has a 60cm wide sponson all round its perimeter, varying in height from 75 to 95cm above the main deck, (*Fig 13*). Its height above the surface of the water varies from 1.2m amidships to 1.68m at the stern and 3m at the vessel's clipper bow. A sprinkling system, with nozzles spaced 60 to 90cm (2 to 3ft) apart is fixed on the fishing platform along the stern, side and bow provides the required spray during fishing. The wooden chum tank, which is 89 × 61 × 63.5cm (35 × 24 × 25in) is located on the port side of the raised deck near the stern.

3.4 Hawaiian sampans

The Hawaiian skipjack pole-and-line fishing boats have evolved from the Japanese tuna boats introduced into that part of the Pacific Ocean around the turn of the century. Since those early days, modifications in design to meet local conditions and preferences have resulted in the emergence of a distinct type of vessel (*Fig 14*) which also incorporates some features of the American tuna clipper. The modern sampans have a sponson similar to Japanese boats but only along the sides of the after deck and across the stern. This sponson is 1 to 1.25m wide and is raised up to 60cm above the deck. Chumming and angling take place, as in the tuna clippers, around the stern.

Typical boats are from 15 to 25m overall with a beam of 3.7 to 5.2m. They are generally built of wood with a high, narrow bow and a moderate freeboard aft. There is a pilot house and flying bridge amidships and live-bait wells (usually six) fitted below the main deck, aft of the engine room. The water circulation in the wells is provided either by a series of screened holes along the bottom of the well which allow sea water to enter when the boat is underway, or by a pump circulating system. A method used by the Hawaiian fishermen to provide water circulation when at anchor in calm water is to rock the boat.

The spray system operates through nozzles placed across the stern and along both port and starboard gunwales aft. The boats have no refrigerating system but some of them carry ice. As the fishing proceeds, the bait-wells are emptied and the catch stored in them.

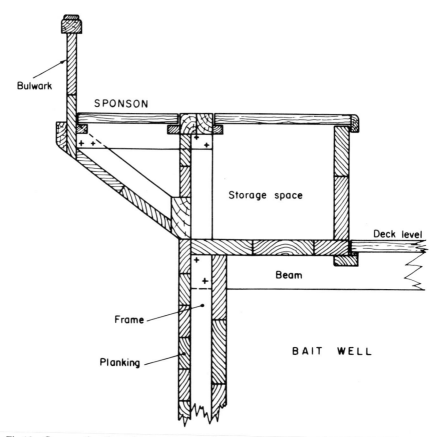

Fig 13   Cross-section through a sponson on a Japanese medium type pole-and-line vessel

3.5   Mini-clippers and other medium-size pole-and-line boats

Medium 12 to 20m specialized boats are used in this fishery in the Azores, Madeira, Cape Verde, Canary Islands, and in the Bay of Biscay. Most of them resemble the tuna clippers in that they have superstructures forwards or amidships and live-bait tanks on the after deck. However, they have no

Fig 14 Hawaiian sampan

Fig 15   Medium size pole-and-line boat from the Atlantic island of Madeira

sponsons and only a few use overhanging fishing racks, the fishing being done from the deck (*Fig 15*). An FAO designed variant of a 25m clipper is shown and described in *Fig 16*.

3.6   Pole-and-line boats for small-scale fishery

Pole-and-line tuna fishing has traditionally been used on a small scale in many areas of the Indian Ocean and of the western and central Pacific. This fishery is supported by 4 to 14m boats ranging from outrigger dugout canoes to western planked-hull type wooden and fibre-glass boats, with either sails or engines or a combination of both.

3.6.1   SMALL JAPANESE POLE-AND-LINE BOATS

The small Japanese type of pole-and-line boats for fishing skipjack, frigate mackerel, small tunas, *etc,* are of 5 to 25grt and diesel-powered. They are equipped with radio navigation instruments, radio telephone, thermometer,

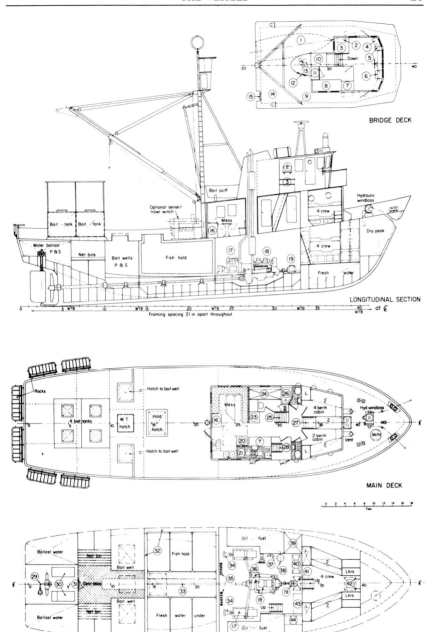

Fig 16 Mini-clipper from Suva Island showing layout of tanks

Fig 17   Small (23t) Japanese type pole-and-line boats

echo sounder, Secchi disc, a sprinkler, a live-bait tank and the usual fishing gear (*Fig 17*). The smallest boats have a crew of one to three fishermen while those of more than 10grt carry a crew of 10 to 15 and have sponsons. As they work in the inshore waters, operating from nearby small ports, their trips are usually of one day, but may sometimes last up to three days. Some carry on pole-and-line fishing and trolling throughout the year, but when conditions make this type of operation difficult they can be used for fishing with handlines, gillnets, or other gear. The small-boat fishermen generally catch their own live bait.

While the gear is similar to that used in medium size vessels, the poles are shorter (3 to 4m). Fibre reinforced plastic (FRP) is now being used to a greater extent than wood for building these boats.

3.6.2 SRI LANKA

The traditional craft used for pole-and-line fishing in Sri Lanka are outrigger canoes. These are more than 4.5m long and are mainly propelled by sail, carrying a crew of five to seven men. However, in the past 20 years mechanized 3½ ton, 8.5m wood or FRP boats have been introduced (*Fig 18*). They are powered by an inboard diesel but instead of bait wells, they generally carry the live bait in towed baskets.

Fig 18 Motorized fishing boat used in Sri Lanka for pole-and-line fishing. Note bait basket alongside

3.6.3 LAKSHADWEEP'S *MAS ODIES*

The *mas odies* is a traditional sailing-cum-rowing skipjack boat of the Lakshadweep, (formerly the Laccadives), now almost entirely replaced by more modern mechanized craft (see *3.6.4*). Its stern is low and the bow rises, ending in a pointed beak (*Fig 19*). The boat is of streamlined design and has a generous keel for sailing to windward. The length overall ranges from 9.75 to 12.2m and the beam from 2.13 to 2.74m. It is plank built with four or five compartments in the bottom for holding live bait. The water enters the middle compartments through holes drilled in the bottom of the hull and circulates to the other compartments through holes in the dividing walls. The live bait are held in the middle sectors and the water bailed out from the end compartments.

The fishing platform, with a breadth equal to the maximum beam of the boat, is built at the stern. There is a wooden post provided near the stern for the helmsman and to nest the mizzen mast when not in use, while the taller foremast carries a single lateen sail. The boats are equipped with 14 to 18 oars, depending on their size and they are manned by a crew of 15 to 20. The boat is rowed during the pole-and-line fishing operation.

Fig 19   A *mas odi* – the traditional pole-and-line fishing boat from the Lakshadweep Islands (Laccadives)

3.6.4   LAKSHADWEEP'S MECHANIZED BOATS

Mechanized vessels were introduced in the 1960s and some of the *mas odies* were also mechanized. The new boats are constructed of wood and built in two sizes, 7.90 and 9.15m overall (*Fig 20*). They have a transom stern and are powered with a 10 to 30hp diesel engine. They carry a 10cm diameter

Fig 20   Laccadives motor boat

mast, 3 to 4m in length, for setting a sail in case of engine failure. The engine is installed amidships and is surrounded by wooden partitions. The space on both sides between the partitions and the hull is planked to form a narrow deck. The engine room is roofed by two planks.

A watertight live-bait tank, 1.6 × 0.8 × 0.8m, is fitted forward of the engine room. A description of this tank and how it functions is given in 6.6.

Fig 21 Artisanal boat for pole-and-line fishing, from the Maldives

3.6.5 MALDIVES' MAS DHOÑIS

There are many types of boats built in the Maldive Islands, those most used for fishing being 4.90 to 6.10m overall, employed in inshore operations, and those of 6.70 to 12.20m used for both inshore and offshore fishing. In the past, all the boats have been built of coconut palm wood but nowadays,

Fig 22   Motorized pole-and-line boat from the Maldives

imported wood from India prevails. They are powered by both sail and oar, (*Fig 21*), and more recently by a 20 to 30hp inboard diesel engine (*Fig 22*). They have a deep bilge which helps to make for smooth sailing, and also serves as a live-bait tank, the water being let in through small plug holes around the mast and bailed out from the fish well aft. The motorized boats have a watertight bulkhead separating the engine compartment from the rest of the boat.

All boats have canoe sterns and the bigger boats have a high stern, balanced by a tiller aft. Thwarts are provided for the rowers, the big boats using four pairs of 3.60m oars (*Fig 23*). The boats are propelled partly by rowing and partly by a sail of 30sq m area. A bridge is set up around the mast, which is placed about a third of the length of the boat from the stern, and here the crew can sit for their meals. Anchors (made of wood) and scoops, fishing tackle, *etc* are kept in the fore end of the boat. The rudder is tied to the stern post by a rope.

Fig 23 Motorized pole-and-line boat from the Maldives, showing thwarts, oar crutches and boom crutch at the stern

The boats are saturated with shark oil when they are built, as much as 40 litres being used for the bigger boats. Every second week the boats are beached in order to coat the underwater area again with shark oil. This is done not only to preserve the wood but also to improve the boats' speed. The boats have a useful life of about 30 years.

3.6.6 TAHITI FAST BOATS FOR SHELL-LURE ALBACORE FISHERY

There is a very marked contrast between the old and new styles of tuna fishing in the Tahiti area. In the outlying atolls of the Tuamotu Archipelago the artisanal fishermen still use the traditional one-man outrigger canoes. They launch them through the surf and paddle a few hundred metres offshore to catch skipjack.

However, many of the fishermen are now using powerful, fast and expensive boats. These are 10 to 11m overall, and are strong, wooden-built launches. They are powered by 250 to 300hp diesels and can operate at speeds of up to 25 knots. While there are a few older boats in service with engines of less than 100hp, some 80 per cent of the present tuna fleet can operate at 16 knots or more, the latest ones, as described, reaching 25 knots. Such high speed boats are costly to construct—about US$30,000 in 1975, including engine—and also in maintenance. This expense is justified by the need to operate at high speeds for fishing and for rushing the catch back to harbour. The local market has a strong preference for freshly caught un-iced skipjack and the first boats back get the highest price for their catch,

Fig 24   A pearl-shell lure tuna fishing boat from Tahiti

while the last ones may have difficulty in selling theirs. The relatively high price fetched by fresh tuna makes the operation of these boats profitable.

The design of the boats is that of a hard-chine, fast day cruiser, built to withstand the strain imposed by their engines and the pounding to which they are subjected when travelling at high speed through choppy seas, (*Figs 24* and *25*). Although FRP construction has been tried, wooden construction remains standard, the hull being planked on wooden frames, and the deck and cabin made of marine plywood. The cockpit is always made to be self-draining. The steering wheel is fitted aft of the cabin near the console but all boats have a tiller for use during fishing operations, (*Fig 24*).

A removable rack for stacking fish is fitted in the cockpit and the fishing poles are carried in special fittings on the roof of the cabin (*Fig 25*). As all the boats return to port each night, their cooking and accommodation facilities are rudimentary. Most of the boats are now equipped with two-way radio. The boats are usually manned by a crew of two fishermen and a 'boy' or apprentice who removes fish from the lines and clubs, bleeds and stacks them in the racks while the fishermen make the catch.

3.6.7   NEW DESIGNS

Two small pole-and-line boats have been designed by FAO, one for inshore 'day fishing' and the other faster and able to go further offshore. Each has two insulated holds, one for the live bait and one for the catch. The live-bait well can be used for the catch at the end of the fishing day. Two to five sleeping berths and a superstructure to contain a cooker, *etc* can be provided. The fisherman may join forces with other boats, if necessary, to catch live bait using liftnets, *Figs 26* and *27*.

Fig 25   A Tahitian pearl-shell lure fishing boat. Note the semi-displacement hull and the racks for fishing poles on the cabin top

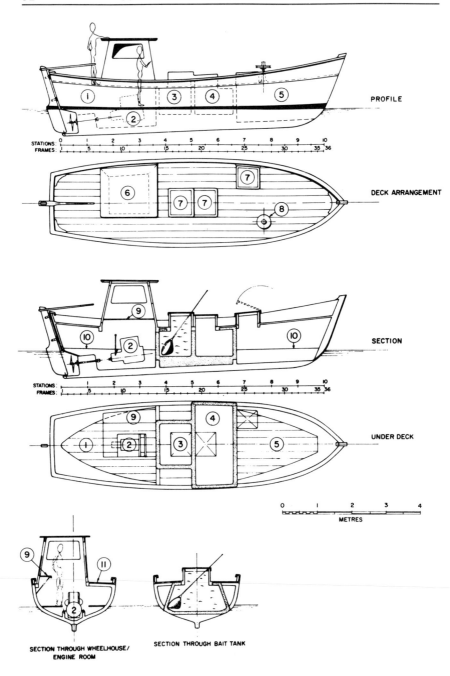

Fig 26   FAO design for a 'day-fishing' inshore pole-and-line and bait fishing boat

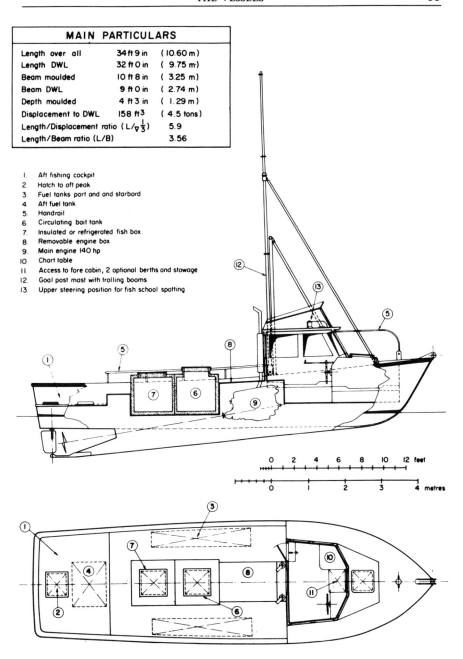

MAIN PARTICULARS

Length over all	34 ft 9 in	(10.60 m)
Length DWL	32 ft 0 in	(9.75 m)
Beam moulded	10 ft 8 in	(3.25 m)
Beam DWL	9 ft 0 in	(2.74 m)
Depth moulded	4 ft 3 in	(1.29 m)
Displacement to DWL	158 ft³	(4.5 tons)
Length/Displacement ratio ($L/_{\nabla}\frac{1}{3}$)	5.9	
Length/Beam ratio (L/B)	3.56	

1. Aft fishing cockpit
2. Hatch to aft peak
3. Fuel tanks port and and starbord
4. Aft fuel tank
5. Handrail
6. Circulating bait tank
7. Insulated or refrigerated fish box
8. Removable engine box
9. Main engine 140 hp
10. Chart table
11. Access to fore cabin, 2 optional berths and stowage
12. Goal post mast with trolling booms
13. Upper steering position for fish school spotting

Fig 27   FAO designed fast, small-scale, pole-and-line fishing boat with space for two sleeping berths

CHAPTER 4

THE CREW

4.1 General

The crew of a tuna pole-and-line fishing vessel must, of course, be determined by the size of the vessel. As the descriptions of vessels have shown, their size range is extensive, from the one-man outrigger canoe to the big distant-water vessel which may have a crew of as many as 45 men.

The number of crew depends on several factors of which the most important are:

1 The size of the vessel and the number of men it can carry and sleep.
2 The duration of the fishing trip; on short, one-day fishing trips more men can be carried by the same vessel than on long trips.
3 The economic feasibility of the operation and manpower availability.
4 The degree of mechanization; (the introduction of automatic angling machines revolutionizes the fishery where their introduction is feasible and where suitable vessels and investment capital are available).

Whatever the size of boat and crew, one thing always remains true: the fishing results depend in the highest degree on skill and experience (both general and local) of the fishermen. This should be borne in mind wherever tuna fishing with pole and line is being undertaken by inexperienced personnel, for in such a case the results obtained would not necessarily reflect the actual yield prospects of the grounds and stocks fished.

4.2 The crew of Japanese skipjack vessels

A description of the jobs to be done and the skills required from the crews of the ocean-going vessels in Japan will serve as a general guide for the manning of larger boats and vessels.

4.2.1 THE SKIPPER AND THE MATE

These must be skilled in navigation and seamanship, have knowledge of oceanography and meteorology and be familiar with marine laws and regulations. In addition, the skipper and the mate should have practical experience in handling, maintaining and repairing navigational and hydro-acoustic equipment and all fishing gear and equipment.

4.2.2   THE CHIEF FISHERMAN

One of the key men of the crew, because he is responsible for selecting the grounds to be fished, and for supervising the fishing operations. Thus, he should have had long experience in such fishing and a very good knowledge of fishing grounds and fish behaviour. He should have a permanent assistant to work with him, capable of taking over the duties of the chief fisherman in an emergency. In recent years one man, a *masterfisherman*, has often been appointed to perform the duties of both skipper and chief fisherman.

4.2.3   THE BOATSWAIN

A man who not only has a wide range of practical knowledge and experience in fishing but also has the quality of leadership. He is the senior deckhand who, under the direction of the skipper or chief fisherman, is responsible for supervising the work of the other deckhands.

4.2.4   THE FISHMASTER

The man largely responsible for the quality of the catch and its products. He is in charge of freezing, refrigeration and storage of the catch; of preparing the freezing brine, controlling the temperature and so on. If he slips up in his work, part or all of a catch could be damaged or even completely spoiled. He must, therefore, have sound knowledge of the refrigeration and freezing system, machines and equipment and have had extensive experience in handling and storing fish on board.

4.2.5   THE LEADING FISHERMAN

This is the man who stands first in the bow of a tuna vessel. The man in the next forward position is known as the second bow fisherman, then come numbers three, four and so on. As the front positions on the bow fishing platform are considered the best for the fishing operation, they are allocated to the most experienced and skilled fishermen.

4.2.6   THE CHUMMERS

The men who have the job of chumming, *ie* tossing the live bait into the sea to attract the tuna and keep them swimming close to the vessel. One is assigned to each live-bait chumming tank. The success of the fishing operation depends to a large extent on the chumming skill of these fishermen, a skill acquired through long experience.

4.2.7   THE ENGINEERS

The vessel's engines and all other machinery on board must be kept in full working order, otherwise fishing is interrupted or impossible and any catch on board could be damaged or completely spoiled. Maintaining the main and auxiliary engines and all other electrical and electronic machines is the responsibility of the chief engineer and his assistant engineer. The engine room staff is supervised by the 'chief motorman' under the instruction of the chief engineer or his assistant.

4.2.8   THE RADIO OPERATOR

The man who collects and exchanges information on oceanographic, weather and fishing conditions. He prepares the situation maps and synopses and conducts ship-to-shore and ship-to-ship communications. He needs to be fully trained and experienced not only in these duties but also in the practical work of keeping all his instruments and equipment in working order.

4.2.9   THE COOK

A professional cook is required for the proper feeding of the crew to keep them fit and strong. This applies to vessels engaged in fishing in distant waters but not, of course, to inshore boats which are out fishing for less than a day. Even offshore vessels at sea for two or three days may not need a professional cook, as the crew can take turns in preparing meals.

4.3   The problem of command in new fisheries

A modern pole-and-line fishery may develop from an artisanal fishery by the introduction of motorized boats and vessels able to sail to offshore and–or more distant waters, ie to fishing grounds not accessible to the artisanal boats. In such cases, both regulations and common sense require that the new vessels be commanded by skippers with good knowledge of navigation and seamanship. This involves operation of electronic equipment, knowledge of international emergency procedure and many other skills requiring a rather high literacy level. In countries where such knowledge and skills do not exist together with practical fishing experience, the responsibilities of command should be divided between the skipper, who is the master of the vessel and has an overall responsibility over the crew and vessel's safety, navigation discipline and economy, and the chief fisherman— an experienced and respected practical fisherman—who would be the skipper's adviser and who takes command of the fishing operation. This would be similar to the Japanese practice described above (*4.2.1* and *4.2.2*). After some years, with the appearance of literate fishermen on the local scene, both functions can be integrated in that of a 'masterfisherman'.

CHAPTER 5

LIVE-BAIT FISHING TECHNIQUES

5.1 Physical conditions affect the selection of technique

The collection of live bait for tuna fishing is influenced by many factors, such as the fishermen engaged in it, the conditions under which the fishing for bait has to be conducted and the technical level of the fishery. There are also, of course, economic and political factors which must be taken into account.

Fishermen engaged in fishing for bait in any country will be well aware of the fishing conditions affecting their operations, such as the species and habitat of the available bait fish, how they behave, their survival rate during and after capture, and so on. For instance, where the live-bait schools are found above smooth but not too muddy bottoms, they will know that the use of small beach seines may provide the best means for catching the fish, so long as they can be rounded up, concentrated and collected without damage. On the other hand, if the fish congregate over shallow rocky bottoms or among coral reefs, drive-in set nets or liftnets may be the only suitable gear for catching them. Alternatively, if the fish in such rough bottom areas respond to light, a light-boat may be used to attract and guide them slowly to where a surrounding net or boat seine can catch them. However, if the live-bait schools are found in deep water areas, the only effective way to catch them is by surrounding nets or liftnets.

As already stated, the technical level of the fishery affects the fishing operation. For example, in some technically advanced fisheries the catching of live bait has reached an industrial level, so that a live-bait fishing industry exists to catch and supply bait to the tuna fishing fleets. Such an industry uses larger vessels, equipped with winches, auxiliary motor boats, generators and other necessary modern gear and equipment.

This level of technology is not required in small-scale bait fishing, which is carried out by some tuna pole-and-line fishermen as a part of their overall operation. Even so, they can, of course, make use of similar gear of small size, such as small liftnets, surrounding nets, beach seines and drive-in nets, as well as electric, kerosene and gas lamps for fishing with light.

5.2 Economic and political factors

The economic and political factors affecting live-bait fishing have tended to become more important in recent years because of inflation, rising costs, the extension of national sovereignty over increasingly wider areas and the 200 mile EEZ (Exclusive Economic Zone). Thus, the use of larger vessels in pole-and-line fishing in more distant waters has resulted in the industrial type of live-bait fishing referred to above. In Japan, for example, live-bait fishing has become a separate occupation, using specialized vessels, gear, personnel and techniques. This method of supplying the live bait is economic because it allows the large, ocean-going tuna vessels to concentrate on the main fishing operation without being dependent on bait supply from the shores and waters of other countries. Of course, both small- and medium-scale tuna fishing are also conducted by Japanese tuna fishermen catching their own live bait.

Political factors, such as fishing rights, including traditional rights, may restrict live-bait fishing, or stocks may exist in areas in dispute between countries, where claims of jurisdiction may overlap. In such cases some pole-and-line fishermen may find that they either have to seek their live bait in more distant areas or allow for the supply of bait from foreign national waters. Such changed conditions may require the introduction of different gear and different methods of fishing, and may, of course, affect the whole fishing operation. The new 200-mile EEZ has imposed severe constraints on some of the existing pole-and-line fisheries, mainly because of the live-bait supply problem, but, on the other hand, it has created favourable conditions for the development of small-scale, inshore fisheries for both live bait and tuna, where formerly only large-scale, industrial and mostly foreign vessels were to be seen.

5.3   Nets used for live-bait fishing

The nets most commonly used for live-bait fishing include seines operated from the beach or from boats; surrounding or roundhaul nets; lampara

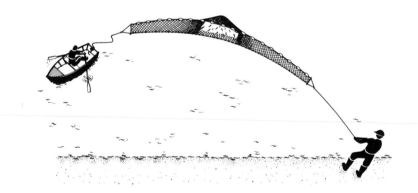

Fig 28   A small 2-man beach seine. Many seines of this type and size are used throughout the world for catching fry and live bait

seines, purse seines and ringnets; drive-in nets and liftnets, including stick-held dipnets and blanket nets. There are variations of each of these nets which have been evolved to meet the particular conditions in a country or a region.

5.3.1 SEINES

The seine is a long-winged net which may or may not have a bag, usually placed at its centre. Those operated from the shore are known as beach, shore or drag seines (*Fig 28*), because they are dragged over the sea bottom to the shore. Those operated from a boat are known as boat, or bottom, or

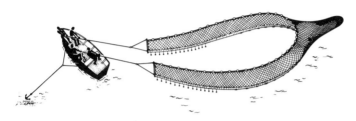

Fig 29   A boat seine

Danish seines (*Fig 29*), and are sometimes brought afloat during the fishing operation. Seines are usually fitted with a rope on each wing, and these enlarge the area from which the fish are herded during the fishing operation. Small versions of the several types of seines are widely used in live-bait fishing.

5.3.1.1 *Beach seines*

While all types of seines have much in common in their design, each has some special feature. Thus, a common feature of beach seines is usually the decrease in mesh size from the wings toward the bunt or bag. The larger mesh at the wing tips, usually bigger than the fish to be caught, has the function of scaring the fish toward the smaller mesh part of the net rather than preventing their escape. By the time the crowded fish may attempt to get away, most of the seine will have been hauled in or 'dried up' so that only the small mesh part will be left in the water.

However, in fishing for live bait or other small fish, such as fry for stocking, short beach seines with a uniform small mesh are sometimes preferred, as are small bagless seines, because it is easier to scoop the fish from them. While beach seining is usually carried out in daylight, there is some fishing done at night; for example, in certain districts of Japan, where the fishermen use underwater lamps to concentrate the fish and lead them to a place where a net can be set.

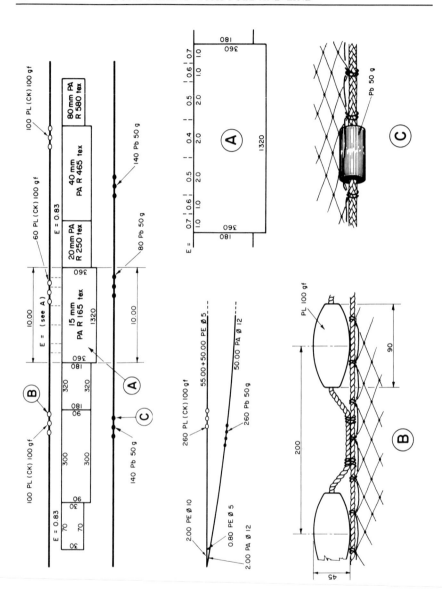

Fig 30   A 50m beach seine for live-bait and fry fishing. The bunt in the centre of the net is created by incremental slack

Two designs of seines suitable for live-bait fishing are shown in *Figs 30* and *31*.

An example of a beach seine operation is provided by skipjack fishermen in Hawaii. They use a Kuralon net, 73m (240ft) in length with a depth of 4m (13ft) at the centre, tapering to 2.7m (9ft) at the ends. The mesh size is 3.5mm. The head rope is supported by synthetic floats and the foot rope is fitted with lead weights (*Fig 31*).

The fishing is usually done in shallow water up to a depth of about 1.83m (6ft), a shallow draught skiff, fitted with an outboard, being used for carrying and setting the net. One end of the net is held by fishermen on the beach while the skiff encircles the school of fish, paying out the net. The outboard is used in this operation if the water is deep enough. As soon as the net is paid out the two ends are drawn together by fishermen who stand inside the encircled area and keep the lead lines close to the ground to prevent the escape of the fish during the hauling operation. Several fishermen do the hauling while others, wearing goggles, dive to ensure that the lead lines are not snagged and are brought together. When the hauling

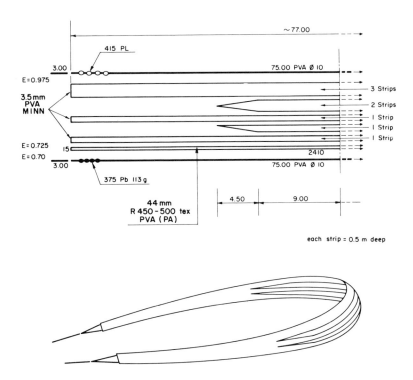

Fig 31   A 75m Hawaiian live-bait beach seine of square 'minnow' netting. Here the bunt is created by incremental depth in the centre, achieved through wedge-like inserts

concentrates the fish in a small section of the net, the catch is transferred to a bait receiver. This is done by placing the cork line inside the entrance of the tilted receiver and raising the webbing of the net slowly and evenly so that the fish swim into the receiver without being touched by hand.

5.3.1.2 *Boat seines*

Although boat seines can be used in live-bait fishing, they are not favoured as they cause a lot of damage to the fish in the dragging operation, so that the catch suffers a high rate of mortality. In view of this, it is recommended that other methods of catching live bait should be used wherever possible.

Most boat seines have bags. They are relatively short nets and are set from one or a pair of boats (*Fig 29*). When the fish are encircled the nets are hauled on board, the catch being 'dried up' at the side of the vessel. The nets, which have sweeplines, are set in a circle or a square and surround a large area. The hauling operation can be performed with the boats at anchor or under way. In some cases, such as those of the Japanese sardine dragnet *bachi ami* and the sardine-silverside seine, the net is towed in midwater at a speed of about one knot for as much as two hours before it is hauled in. These floating seines are used in fishing for relatively slow swimming surface species such as anchovies, silversides, sand lance, smelts, *etc*.

The sardine-silverside seine is bagged and has two sweeplines. The 15 to 18m long bag is about 6m in depth, cylindrical or conical in shape and made of 9mm stretch mesh (or 4¼mm square) minnow netting. The wings of the net are each about 50m in length, the main part being made of 9 to 10cm stretch mesh. Shoulders of 3 to 5m in length connect the bag to the wings. Floats, sinkers and a wooden spreader are placed at the wing tips, the spreaders being connected by a 5m chain to the sweeplines to prevent the net mouth 'swimming up'. A marker buoy is fitted at the centre of the floatline.

5.3.2 SURROUNDING NETS

These are also known as roundhaul nets. A common feature of all types of surrounding nets for live-bait fishing is the small mesh used, if not throughout the net, at least in the bunt. The stretched mesh size ranges from 8 to 20mm, nowadays made mainly from synthetic twines of about R100 tex (210D × 4) to R300 tex (210D × 12). The small size of the mesh and the relatively thick twine used in making the bag prevent the fish being gilled.

5.3.2.1 *Lampara nets*

At one time, before purse seines became more popular for this operation, the American pole-and-line fishermen of the tuna clippers made considerable use of lampara seines for live-bait fishing. A typical operation was carried

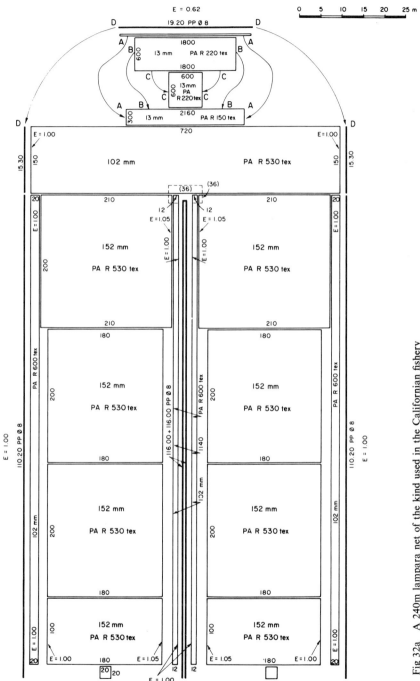

Fig 32a   A 240m lampara net of the kind used in the Californian fishery

out by three boats to locate and capture the bait—a powered boat capable of 30 knots, a net skiff and a 'dry boat'. The powered boats were about 5 to 5.5m overall with a 1.2 to 1.5m beam, fitted with a depth sounder to help locate the fish and powered by a high speed petrol engine of 100 to 150hp. Some of the boats were fitted with reduction gears which permitted the use of propellers of larger diameter to reduce slippage when towing the other boats.

The net boats were of much the same length as the powered boats, with a beam of 1.8 to 2.2m and a depth of 51 to 76cm. The dry boats were flat-bottom skiffs of about 3.7m in length. They were used to hold the end of the net in the setting operation and to support the bag for the catch. Six to eight men were required for setting and hauling the net while two men crewed the powered boat—one to steer and the other to search for live bait and direct the fishing operation. When not in use, the boats were carried aboard the tuna clipper.

When a school of bait fish was located, the dry boat was used for holding the standing end of the net while the powered boat towed the net skiff to encircle the fish. If, by mischance, the length of the circle turned out to be more than the length of the net, a running line was paid out to enable both ends of the net to be pulled to the skiff. Another critical point was in shooting the bag because the lead line of a lampara net is much shorter than the float line. The bag, therefore, had to be so piled that it could be set as a unit to avoid the danger of being torn. The net was hauled in from both ends, with one man repeatedly throwing a weighted line into the space between the wings of the net to scare the fish so as to prevent their escape. During this operation the dry boat picked up the cork line at the bag to give support (see *Fig 32b*).

When the hauling-in was completed, the fish could be transferred directly to the tuna clipper if the water was deep enough to permit this operation. If not, the catch was put into a floating cage and towed to the clipper and the fish transferred with scoop nets made of 12mm mesh and holding 4 to 7kg of fish.

The lampara nets vary much in size. The bigger of the hand-operated lamparas range from 238 to 293m in length and from 15.25 to 23m in depth at the bag. The mesh sizes in the wings vary from 125 to 200mm stretch mesh, gradually becoming smaller until they are only 13mm stretch mesh in the bag. The nets are lightly floated and ballasted in the wings, having about one cork and one lead every 30 to 45cm, but at 15cm intervals at the bag. The bottom line along the bag is solidly ballasted with lead sinkers. *Figs 32a* and *b* show a 240m net used in the Californian fishery.

Smaller, fine-mesh lamparas are used for live-bait fishing around rocky shores. The nets range from 38 to 46m in length and 6 to 10m in depth. They are rigged with 6 or 7 corks and leads per metre and have a mesh size of 20 to 25mm stretched. They are used as a surrounding net, stretching from the surface to the bottom, even if it is rocky. In working over a rough bottom it may be necessary to use divers to ensure that the net is kept free

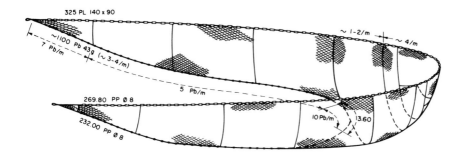

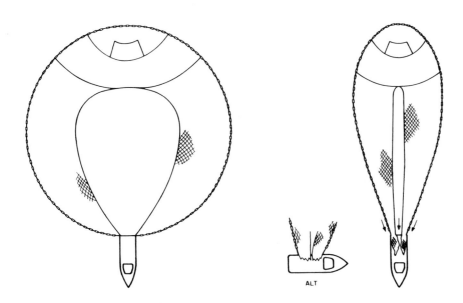

Fig 32b   Operation of 240m lampara net in the Californian fishery

and to keep the lead lines together in hauling. As this underwater work requires much bending over the net, SCUBA gear is preferred to helmets with air supply from the surface. Lampara nets, of course, can also be used effectively in fishing with light.

5.3.2.2. *Purse seines and ringnets*

Both these types of seine can be closed completely under the fish. The former have the bunt at one end and are hauled-in by one wing and are designed for one-boat operations. The ringnets are hauled-in by both wings simultaneously, have the bag in the centre and are often used in two-boat fishing. Both types of nets, made in small size mesh, are useful for live-bait fishing. However, attention must be paid to the question of mesh size, since it may be desirable to use a smaller size than the 20mm stretched mesh recommended in some designs. The choice of mesh size should be determined by the size of the live bait to be fished. The overall size of the net and its design depend on the size of the fishing boats, their level of mechanization and the local fishing conditions.

In *Figs 33, 34, 35, 36* and *37* some purse seine designs are shown which are suitable for live-bait fishing. The designs in *Figs 33* to *35* should preferably be mechanically hauled, but they are still small enough to be hauled manually by a crew of 6 to 10 men. Pursing must be carried out mechanically.

The sardinella purse seine (*Fig 33*) is used mainly for night fishing with light attraction over smooth bottoms of 20 to 70m depth. The Israeli purse seine from Lake Kinneret is used successfully for both day and night fishing (*Fig 34*) while the Italian anchovy purse seine is used mainly for day fishing (*Fig 35*). The first two nets are fished from 10 to 12m boats, the latter from 15 to 20m boats. All three might be suitable for commercial scale operation aimed at the supply of live bait to tuna fishermen. *Fig 36* shows a design for manual operation throughout and is a seine used by French pole-and-line fishermen working from auxiliary skiffs in the Eastern Atlantic.

The purse seine illustrated in *Fig 37*, although larger than the previous net, can be operated manually and can be used for both day and night fishing. Although it is small, it can be used by vessels of up to 20m as well as by smaller ones.

A ringnet evolved for catching small fish in Lake Tanganyika (*Fig 38*) can be used for bait fishing with two boats. It is a suitable gear for vessels of 10 to 15m, with engines ranging from 40 to 120hp. The net has to be pursed mechanically but can be manually hauled by 8 to 12 men. Where the bait is found in shallow water, the lower 40mm mesh panel of the net can be omitted and the leadline, with its 6-mesh wide skirt, joined directly to the 12mm main netting. This ringnet is also suitable for fishing on a commercial live-bait supply scale.

Figs 39 and *40* provide designs of a purse seine and a half-ringnet for live-bait fishing by smaller vessels—those up to and about 10 to 12m in

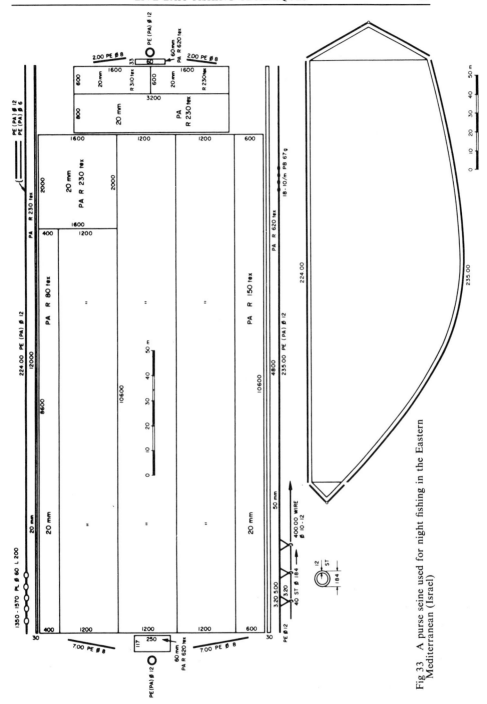

Fig 33 A purse seine used for night fishing in the Eastern Mediterranean (Israel)

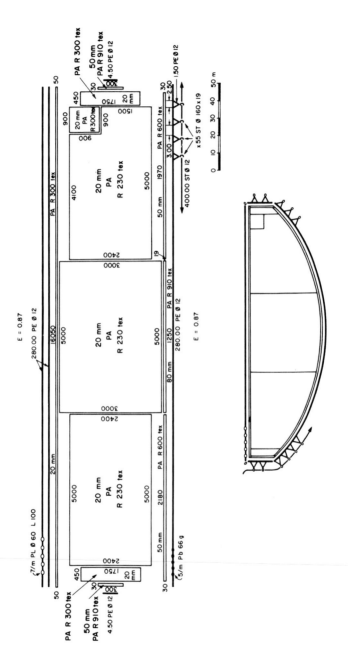

Fig 34 A purse seine used for light-fishing in Lake Kinneret (Sea of Galilee) for day and night fishing

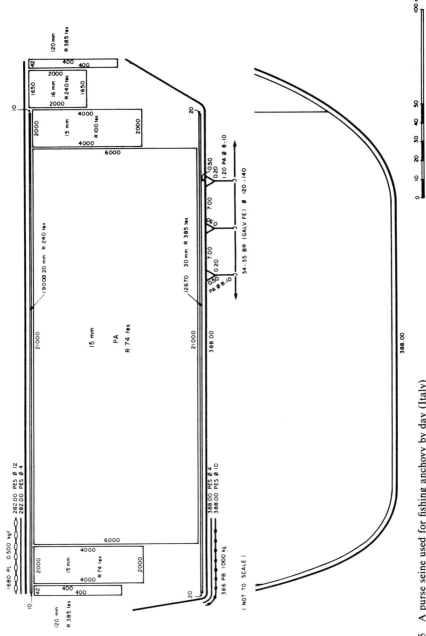

Fig 35   A purse seine used for fishing anchovy by day (Italy)

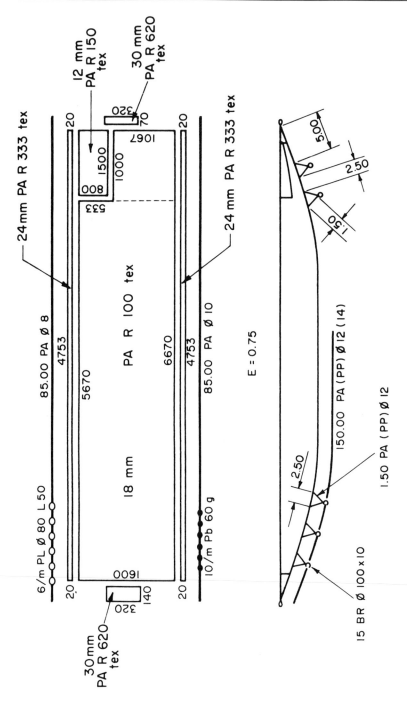

Fig 36 A small French live-bait purse seine for use from auxiliary skiffs (Courtesy Ets L le Drézen, France)

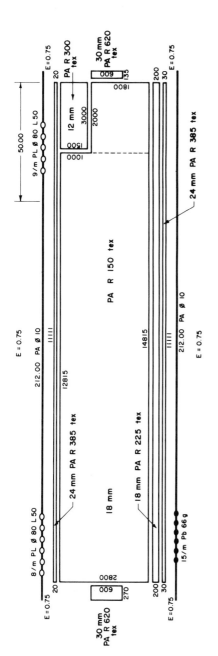

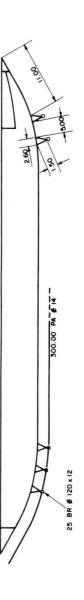

Fig 37   A live-bait purse seine used by French tuna clippers (Courtesy Ets L le Drézen, France)

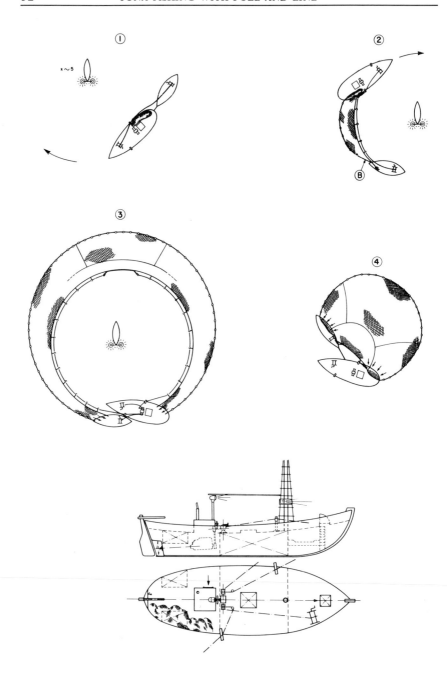

Fig 38a   Operation of two-boat ringnet from Lake Tanganyika (Burundi)

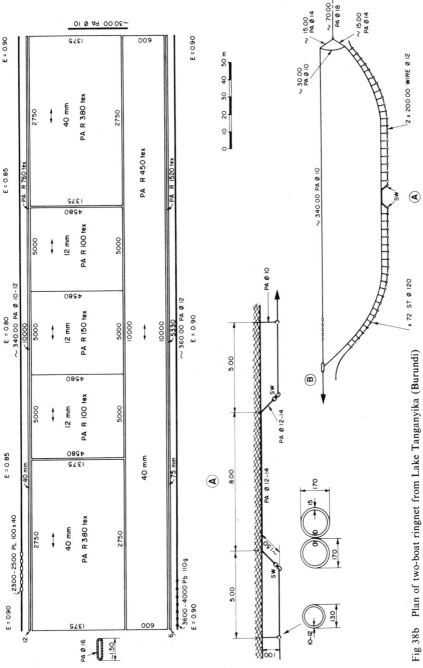

Fig 38b   Plan of two-boat ringnet from Lake Tanganyika (Burundi)

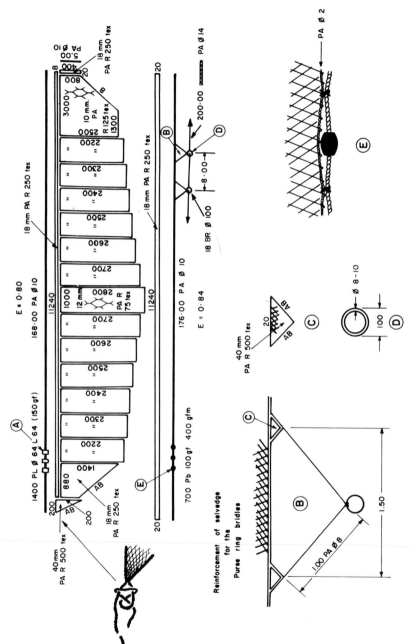

Fig 39   An FAO designed purse seine for smaller boats of 10–12m length

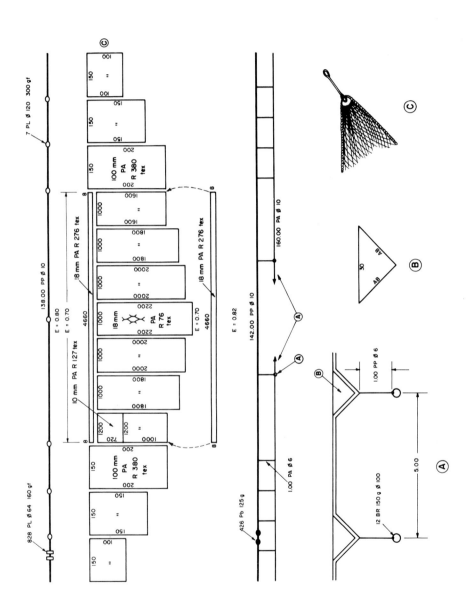

Fig 40   Half ringnet for fishing live bait from one or two small boats

length. The net shown in *Fig 39* requires mechanical pursing while that in *Fig 40* can be successfully operated manually. It should be pointed out that boats of below about 9m in length can operate with these nets in calm water but under less favourable conditions they would have to use smaller and lighter nets.

5.3.2.3 *Chiromila net*

The new *chiromila* pursed semi-surrounding net is an improved version of the original net designed and used by the fishermen of Lake Malawi. FAO gear specialists adapted it for light fishing in Lake Tanganyika (*Fig 41*). The new net is semi-circular in shape, with the floatline attached to the straight edge while the leadline, which is fitted with purse rings and a purse line, runs along the semi-circular edge. *Fig 42* shows the operation of the net, which calls for the use of a motorized boat of about 9m in length, equipped with a line hauler, plus a lightboat and an auxiliary canoe.

It is suggested that this improved net could be adapted for live bait-fishing where a short but deep fishing net is required.

5.3.3 Drive-in nets

The usual shape of a drive-in net is that of a dustpan, the depth of its walls depending on the depth of the water to be fished. Where the water is shallow enough for wading—say, up to about 1.5m in depth—the fishermen can walk-up the fish and drive them into the net (*Fig 43*). In deeper water this has to be done by fishermen diving and swimming or by boats (*Fig 44*). In the latter case the boats move slowly toward the net with the fishermen beating the water with poles or paddles to drive the fish into the net. The fish can also be herded in this way by the use of scare lines *(Fig 43)*.

Drive-in nets can also be used for fishing with light at night. In this operation a light station is set up in the vicinity of the net. When sufficient fish have been attracted, the light buoy (*Fig 45*) is towed or the lightboat is rowed slowly over the net. When the fish reach the net and are in it or over it, the bottom line is lifted across the net opening and the fish are trapped.

Normally, live-bait fishermen set their drive-in nets close to the shore where they have located the school of fish, usually in a shaded area or near to rocks where the fish find shelter and can feed. In setting the net, care must be taken to ensure that it is not carried away by a current. To avoid this, the net is usually held in position by stakes driven into the sea bed (*Fig 44*).

When the fish have been herded into the net, the lead line, which keeps the mouth open, is lifted and hung on the stakes, thus closing the net. If the net is not set deep enough the fish may be crowded and start to jump out of the net during this operation. The fish can be driven into the bunt by lifting the bottom of the net and from there they can be transferred to a floating pen or cage.

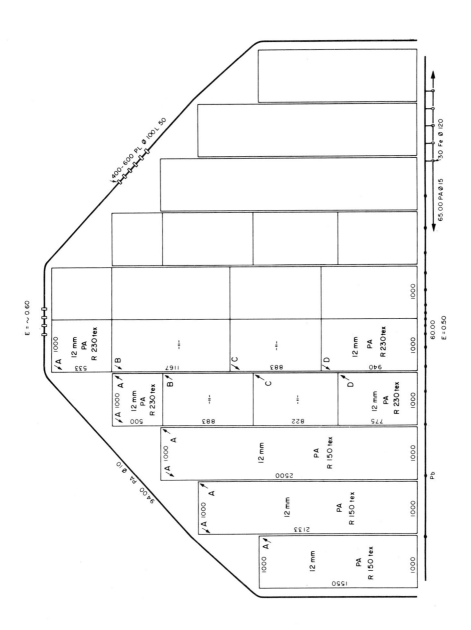

Fig 41 *Chiromila* net

There are variations in the fishing techniques used with this net. For example, cardinal fish (*Apogonidae*) hide in crevices during the day and have to be flushed out and driven into the net. The fishermen do this by thrusting twigs and leaves into the crevices. They also use a special method

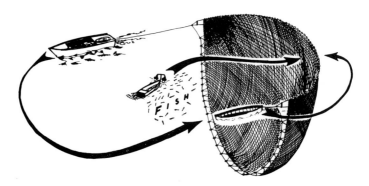

Fig 42 Operation of the *chiromila* net, from Lake Tanganyika. After passing the wing end over to the main boat, the canoe moves to the centre of the corkline

for catching the fish on their return from feeding at night. The fishermen set their nets over the reef at sunset but in such a way that the fish can move out freely at night, as they usually do. Then, when fish return to their habitat early in the morning, the fishermen are there, waiting to drive them into the nets.

Fig 46 is a design of a dustpan shaped drive-in net based on an illustration of a Philippine variant. A similar but wider version of drive-in net (20m across the mouth) is used by some Japanese pole-and-line fishermen. Of

Fig 43 A small, drive-in net operation, with scare line

course, the nets can be made smaller and shallower, when required for catching small but not very mobile species, while for bigger and faster swimming fish a net can be fitted with simple wings of different lengths, as indicated in *Fig 44*.

Fig 44 A fixed drive-in net operation

Fig 45 Italian lamp raft

5.3.4 LIFTNETS

Three important advantages in the use of liftnets for live-bait fishing are the ease with which they can be operated, the relatively low cost of the nets, boats and other equipment needed, and the fact that they can be operated

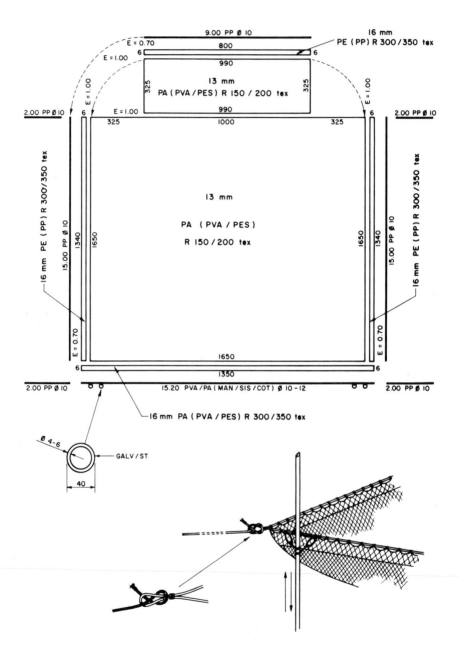

Fig 46   Design of a Philippine drive-in net, showing detail of a corner, with bottom line lifted and the net in working position

over any depth of water and over any type of bottom. Indeed, any tuna pole-and-line fishing vessel can carry a liftnet to catch its own live bait. For these and several other reasons, these nets may well be considered the most important gear for live-bait fishing.

While there are many variations in design, as the examples dealt with here will show, the principle of operation is the same. The net is submerged, the fish are attracted to concentrate above it (by light or by chumming) and the net is hauled up more or less vertically and brought at least partly out of the water to net the fish.

5.3.4.1 Stick-held dipnet

One of the best known liftnets is the Japanese *bouke-ami*, which is a stick-held dipnet (*Fig 47*). This net was first used for catching mackerel and later for fishing with light for saury. Typically, the gear consists of the net, two outrigger poles, one floating pole, and ropes, sinkers, floats, *etc*, but there are variants having only one outrigger and lacking the floating pole. The poles are usually of bamboo. The manual operation of the net requires little in the way of deck arrangements, while almost any kind of available winch can be used for mechanized operation. *Fig 48* illustrates the design and *Fig 49* the operation of a variant of this net as used for live bait in Fiji.

There are many variations in the design of the stick-held dipnet. It may, for instance, look like a box, a bag, a shovel or a dustpan. An example of such a net, used with only one stick, is shown in *Fig 50*. As can be seen, the net is semi-oval in shape, and the edge facing the ship is fitted with rings to enable the net to be pursed. The hauling method to be used determines the rig. *Fig 51* shows the simplest, though large, variant in which the distant edge of the net is hung on a bamboo float boom which is kept parallel to, and at a set distance from, the vessel by two other floating bamboo booms. The sidelines of the net are fitted with floats independent of the booms. The net, with a sinker of 5 to 25kg, is hauled in by means of six or seven lines attached to it. The side-lines, net and booms are hauled in by hand, drying up the net and the catch until the bamboo float boom is brought to the vessel's side so that the fish can be brailed out and onboard.

In another variant, the float boom is replaced by a number of ordinary floats, the distant part of the sidelines being fitted with rings that slide along the outrigger booms. The near part of the net is hauled aboard, then the rest is hauled alongside by ropes which run through blocks at the ends of the booms. In re-setting the net the process is reversed (*Fig 52*).

Figs 49, 51 and *52* illustrate the operation of the stick-held dipnet. Hauling can be improved by replacing the heavy leadline by a positive buoyancy-ballasted line with only two heavy sinkers, one at each corner of the net. With this arrangement the whole length of the buoyant line rises to the surface when hauled and purse rings along the buoyant lower line and the sidelines can be omitted. On the other hand, the use of purse rings, as in the variant in *Fig 53*, enables the whole operation to be carried out by less people.

Fig 47   Two views of a Japanese type of stick-held dipnet

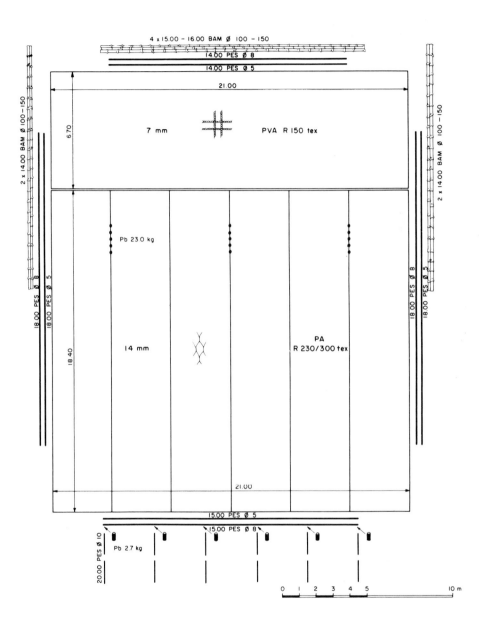

Fig 48   A stick-held dipnet (*bouke-ami*) of the type used in the Fiji Islands

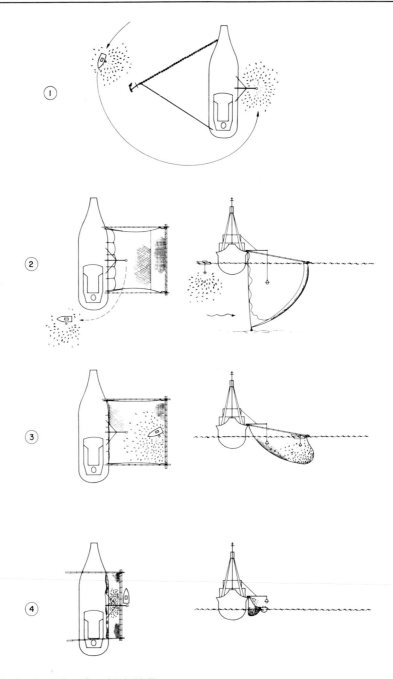

Fig 49   Operation of a stick-held dipnet

As the vessel, net and booms are affected by currents, the relative position of the net and vessel may change. The skipper, therefore, has to be skilful in manoeuvring, in order to maintain correct position. This is usually done by the operation of the engine and rudder, although Japanese vessels often use a spanker sail to maintain the vessel's position crosswise in the current.

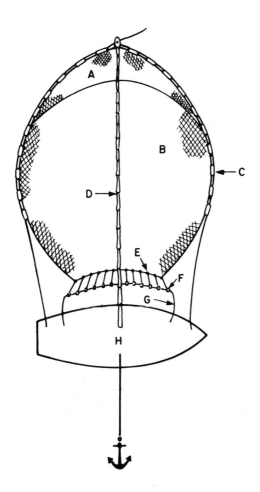

Fig 50   A one-stick dipnet used by smaller boats in Japan

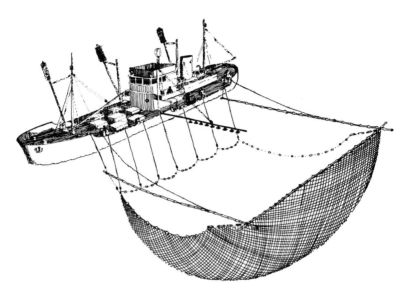

Fig 51   Hauling a stick-held dipnet

5.3.4.2   *Blanket net*

The blanket net is suspended from the side of the vessel and is hauled up to the edges of the outrigger. This is in the opposite direction to that of the stick-held type of net (*Fig 54*). The net, which is formed by a rectangular section of netting varies in size, depending on that of the vessel. A type of net used successfully in the Caribbean for live bait is 16.8m in length and 12.8m in width, the section suspended over the water being made of 75mm stretched mesh and the main netting of 12.5mm stretched mesh. Polyamide twine R tex 450 to 550 is used for the former and R tex 150 for the latter. Fishing is carried out in association with light attraction, preferably in a well sheltered place near to a discharge of fresh water from a river or other source. It can also be used in a harbour, with the vessel secured to a wharf or anchored. The outrigger poles are lowered to be parallel with the water surface, and the net is spread along the side of the vessel and connected to the lines from the outrigger poles. Another pole is used to support an underwater or surface light at a distance of some 3m from the side of the vessel, positioned to shine on the centre area of where the net is to be lowered. It is switched on at dusk. When a sufficient concentration of fish has been attracted the net is lowered so as to hang vertically in the water and, at a signal from the skipper, the crew haul in the lines which have been threaded through blocks at the ends of the outrigger poles and connected

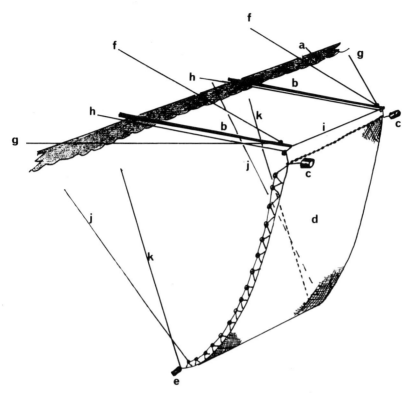

Fig 52   A stick-held dipnet rigged according to a Soviet design.
a.Ship's bulwark; b.Outrigger booms; c.Large floats; d.Net; e.Heavy sinker; f.Topping lifts; g.Guys; h.Floating pull-ropes; i.Span; j.Purse line; k.Leadline pull-ropes. The net is set in working position by pulling the ropes 'h' and letting go the purse line 'j' and the ropes 'k'. Ropes 'h' are rove through blocks fitted at the seaward end of the booms

with the bottom ends of the net. Two men in a row boat often assist in the operation by hauling up the mid-section of the net between the outriggers.

When the top of the net breaks the surface, the side of the net near to the vessel is pulled up until the catch is concentrated in a small pocket. The fish are then transferred by bucket to the live-bait tanks.

A variation of this gear is the framed net, but it can be used only when fishing with fairly large vessels. It has a rectangular bag and is framed by bamboo sticks (*Fig 55*). Long outriggers are needed for its use.

5.3.4.3   *Hawaiian liftnet*

This net, (*Figs 56* and *57*), is normally used when fishing with lights. When setting, the net is paid out from a skiff, one end of the cork line being secured to the port side of the stern of the fishing boat. When the net is fully paid out, the skiff is turned to be parallel to the fishing boat and the

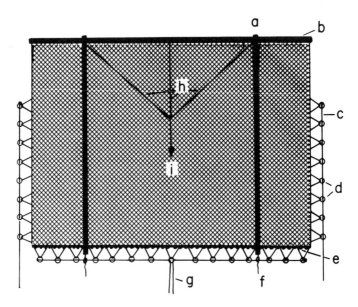

Fig 53   A Russian variant of the Japanese stick-held dipnet. It can be
operated by means of the winch gypsy head or by a capstan.
a.Bamboo stick 10m long; b.Bamboo float 18m long; c.Side purse line;
d.Purse rings; e.Ballast chain, approx 225kg; f.Hand ropes to bring ends of
sticks on board; g.Leadline purse lines, operated by using the warping
drums of a trawl winch; h.Three legged bridle; i.Haul line leading to a
block suspended from the end of the derrick and used to pull the whole net
towards the ship's side

crew pull the lead line and 'rib lines' (haul-in ropes) as the skiff moves
along the outer edge of the net while the fishermen in the boat pull the lead
line from the bow and make it fast along with the 'rib lines'. On the
command of the skipper the net is pulled by the men in both the skiff and
the boat and as it is 'dried up' the catch is pocketed in the bag. From there
the fish are transferred to the bait wells in the fishing boat.

During the fishing operation, the boat lies at anchor over the live-bait
grounds. A lamp is suspended from a pole about 6m from the port side of
the boat and lowered about 1m below the surface. The light is usually
switched on at dusk and left burning until about an hour before dawn, a
rheostat often being used to dim the light just before the net is set. This
tends to concentrate the attracted fish as well as to conceal the movement of
the net in the water.

5.3.5   SMALL BOAT LIFTNETS

The Philippine *basnig* (Fig 58), the Hawaiian *opelu* net, (*opelu* is the local
name for the small scombrids and carangids mainly caught for bait), and
the liftnets used in the Lakshadweep (formerly Laccadive) Islands and Sri

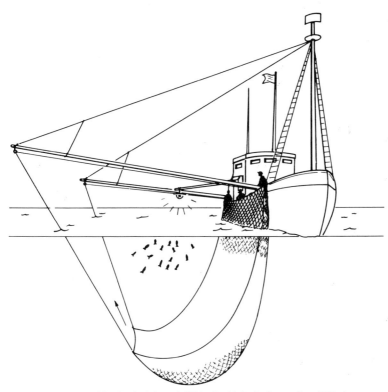

Fig 54 Blanket net used by Radovich and Gibbs for biological sampling (USA)

Fig 55 Framed liftnet

Fig 56   Fishing for live bait, using an underwater light with a Hawaiian liftnet

Lanka can all be used from relatively small boats. The *basnig* can be operated from catamarans and outrigger canoes and the others from ordinary open fishing boats.

5.3.5.1   *The opelu net*

The *opelu* net is arranged on a wooden hoop of about 6m diameter from which hangs a 10 to 11m deep conical bag (*Fig 59*). The netting is usually made of nylon, the mesh size decreasing and the twine size increasing toward the codend. Small lead weights ensure that the net hangs correctly while the hoop is supported by bridles which are joined to a thicker, single, haul rope (*Fig 59*). Hoops are often made from several pieces of seasoned hardwood lashed together to make two rods of about 9 to 9.5m in length. Another way to make the hoops is to use bamboo reinforced with steel rod centres. The rods have to be flexible enough to bend into the circular shape but stiff enough not to collapse when the net is hauled. Spare rods with the lashing already attached to them are usually carried so that repairs, if needed, can be made without delay. The joints between the rods should overlap and be pushed through a rope grommet so that the netting does not catch on them.

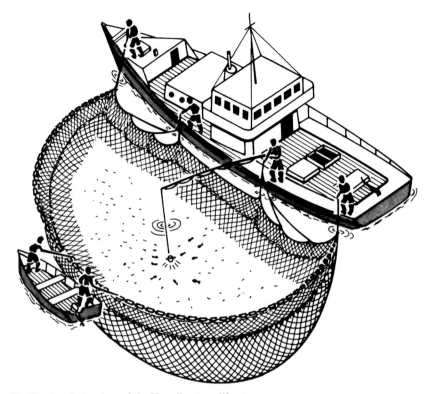

Fig 57   Another variant of the Hawaiian type liftnet

Fishing usually starts as soon as there is enough light to see the fish through a glass bottomed box or bucket. The fish are mostly found in about 10m of clear water. The best catches are made in the early morning and in the late afternoon. The minimum depth for fishing is about 18m, as in shallower depths the codend may become fouled on the coral. The preferred fishing depth is from 30 to 36m or more. The boat is anchored or allowed to drift, depending on wind and current, during the search for the fish.

When a satisfactory school is found, a handful of bait is put in a 30 × 30cm denim cloth weighted at one corner by a lead sinker of about 0.5kg. An 18m thin line is attached to the opposite corner, the cloth is folded over the bait and a slip-hitch is made around the bundle, which is then lowered until it is just above the school. The bait is released by a sharp pull of the line and shakes out in a small cloud and floats down among the fish. White or light coloured bait is best, not only for the fish to see it but also because it provides a background against which the fish are easier to observe. The bait is made from ground fish, from vegetable matter or from both.

When the fishing operation is to start, the codend is tied with the codline which is made fast to a piece of lead. As the net is paid out the rods are

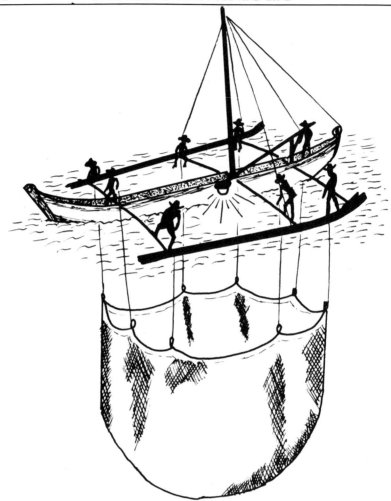

Fig 58   Schematic view of a *basnig* type liftnet operated from an outrigger canoe. It was developed in the Philippines during the 1930s

joined to make a hoop. The end of the haul line is made fast to the boat to keep the net in control. As soon as the net is formed into a circle on the hoop it is lowered, the crew watching to see that it is setting smoothly. The haul rope is paid out until the hoop rests just below the area of the attracted fish and more bait may be scattered to gather more fish over the hoop. Next, the hoop is slowly hauled up while more ground bait is released so that it floats down into the net. As soon as the fish follow the bait into the net, the hoop is hauled to the surface. The two bridles over the joints are taken off and the joints opened, and as the rods straighten, the net is brought aboard and the fish removed from the bag. Two men are needed for this complete operation.

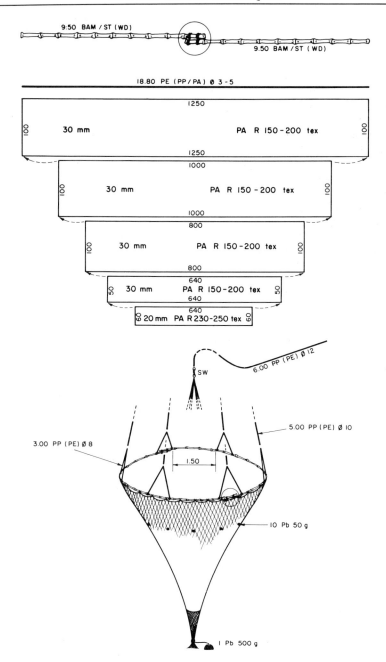

Fig 59 *Opelu* net

5.3.5.2   *The Lakshadweep bait-fish net*

The Lakshadweep bait-fish net is made of a rectangular 4.5 × 5.5m section of 6mm stretched mesh netting of cotton twine (NeC 40/2/3, about 90 tex). The selvedge mesh is made of thicker mesh twine (NeC 40/3/3, about 135 tex). A cotton cord of 3mm diameter is rove through and laced to the selvedge mesh. A Lakshadweep pole-and-line fishing boat carries two or three such nets.

Fishing for live bait is done in the early morning in a lagoon on the way to the tuna grounds. As soon as a school of bait fish is spotted, a fisherman, wearing a diving mask, goes down to see if the school is big enough. If so, the boat gets into position and is held there against wind and current by two anchors. The net is lowered by four fishermen (*Fig 60*) each handling a pole to control a corner of the net. Next, ground bait in the form of a paste made from crab or tuna meat is thrown into the water to float down over the net to attract the fish. Sometimes the bait is spread by putting it on a coir mat which is attached to a pole so that it can be moved up and down over the net, thus spreading bait throughout the water column between the net and the surface. As soon as enough fish have gathered above the net, it is lifted and the catch is transferred to the live-bait tank in the boat.

The Lakshadweep fishermen sometimes use the drive-in method of fishing, which takes place in shallow water. The net is held in position by four fishermen or by pinning one side of the net to the bottom with stones while holding the other side up with two poles to form a dustpan shape. A scare-line made of coir rope, with palm leaves fitted all along its length, is used to herd the fish into the net. The trapped fish are transferred to a floating tank or basket which is towed to the fishing boat.

5.3.5.3   *Sri Lanka liftnet*

The Sri Lanka bait-fish liftnet is some 10 to 12m square and is usually made of hemp (1.5mm twine diameter, 60mm mesh netting). The net has a small mesh square section sewn over the heavier netting at the centre (*Fig 61*) so that the heavy hemp protects the small mesh section when the net is laid over a rocky bottom. However, where the bottom is smooth, the lighter, FAO net (*Fig 62*) may be more effective, having less drag.

The fishing operation requires two vessels, such as two outrigger canoes or one canoe and one mechanized boat. Most of the live bait caught by the Sri Lanka fishermen is redbait and cardinal species, mostly found over rocky bottoms. Bait fishing is carried out in daylight, preferably in the early morning. When a satisfactory school is located, four of the crew handle the net, spreading it over the rock. This action disperses the fish but they soon return. The fishermen then quickly haul up the net by ropes attached to its corners (*Fig 63*) and the catch is transferred to the bait basket secured to the side of the boat. This fishing operation is repeated until at least 20kg of live bait have been caught.

Fig 60   Setting a liftnet in the Lakshadweep (Laccadives)

Fig 61   Sri Lanka square live-bait liftnet

Sometimes certain varieties of live bait rise to the surface where they appear like a red ball. The fishermen take these with a scoop net. Fishermen who are more skilled in catching live bait than tuna often catch the bait for others under an agreement that they will receive 50 percent of the resultant tuna catch.

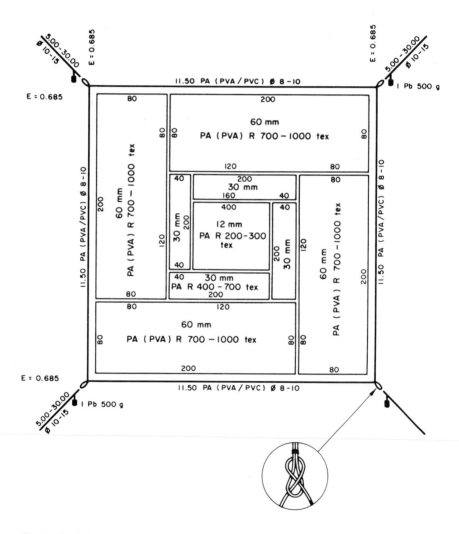

Fig 62   FAO design for a square liftnet for small-scale two-boat live-bait fishing

Fig 63 Two-boat liftnet seen in operation in Sri Lanka

Fig 64 A small, square live-bait net operated by one boat

5.3.5.4   *Other liftnets for baitfish*

An all fine-mesh 7 × 7m net is used in the Maldives. It is spread over the bottom and a paste bait made from bonito meal is scattered through the water over the net. This net, and similar small bait nets are operated from one boat and handled in a way similar to that shown in *Fig 64*.

Fig 65 shows a staked blanket net, known as the *kabyaw* in the Philippines, which under favourable conditions can be used for live-bait fishing.

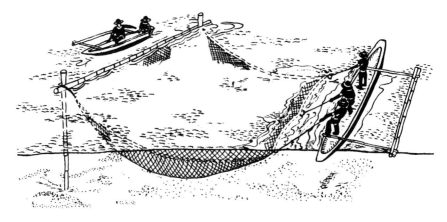

Fig 65   A Philippine staked blanket liftnet, the *kabyaw*

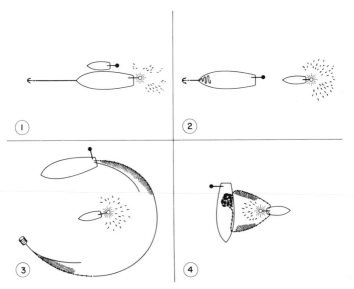

Fig 66   Light-fishing operation using auxiliary skiff and round haul net

5.4   Use of lights

As the foregoing text on fishing for live bait has shown, the use of lights to attract fish plays a considerable part in the operation in many parts of the world. Such lights are used mostly for fishing at night in waters where, during the day, the live bait cannot be located by sight, or where they do not form big enough schools for fishing with surrounding nets.

Fishing with lights is, of course, carried out in many different ways. A recommended method for tuna pole-and-line vessels is for the main vessel to operate the fishing lamp or lamps. Using the light for periods of 30 minutes to 3 hours should be enough to attract a satisfactory concentration of fish. At this point an auxiliary boat (skiff, canoe, *etc*) can take over, using a kerosene or battery operated lamp as the main boat dims its lights and then switches off. The lamp of the auxiliary boat holds the school while the main boat sets the roundhaul net. The auxiliary boat joins in the final stages of hauling and brailing (*Fig 66*).

A lamp of 300W or even less may be enough to maintain a fish concentration. This is often done by an auxiliary skiff with a battery operated lamp while the fishing boat sets the purse seine.

Another method is to install lamps on a raft which is then anchored in the fishing area (*Fig 67*). After the fish have been attracted and have gathered in sufficient numbers, a boat can be used to complete the fishing operation.

As already pointed out, there are various techniques employed in fishing

Fig 67   An Israeli lamp raft

with lights. The choice of method, of course, depends to a large extent on factors such as the level of technology in the locality and the extent of investment in equipment and so on. The lamps used may vary from any type of kerosene, carbide (acetylene) and kitchen gas surface lamps (*Fig 68*) to surface and underwater electric lamps (*Figs 69* and *70*). However, it should be noted that, in general, less light power is required for live-bait fishing than for catching food fish. The cost of obtaining the equipment used in fishing by light attraction for live bait is therefore modest.

Fig 68 Typical fishing lamp, with four incandescent mantles (two are partly covered). The gas burned in the mantles can be either vaporised paraffin (kerosene pressure lamp) or liquid gas (gas lamp). Note that the gallows supporting the lamp can easily be turned round or removed from the steel tube, which is firmly installed on the light-boat

For more information and detailed technical instruction the reader is referred to the FAO Fishing Manual, *Fishing with Light*, listed in the Bibliography.

5.5   Use of bait to attract bait fish

As already mentioned, various kinds of ground baits are used to chum the live bait. For example, many tuna boats working in the Atlantic Ocean use

Fig 69   An electric fishing lamp installed by FAO on board a fishing boat for live-bait fishing

a paste made from ground peanuts and salted cod's roe. This is scattered from a skiff in the vicinity of a school of fish—usually sardines and anchovies. As soon as the fish concentrate on the bait, the fishing boat sets the purse seine and encircles and catches them.

Some fishermen favour the use of ground fish flesh alone as bait, others preferring a vegetable base. Many use a combination, as in the example quoted above. Some believe that live bait caught by using bait of ground fish flesh do not survive as well as those caught with vegetable bait. There is also the fact that fish flesh bait attracts predators and they scatter the schools of bait fish. Thus, in areas where sharks abound it is better to use a vegetable based bait. On the other hand, experience has shown that the small scombrids and carangids feed better on a fish flesh bait. The best answer to such situations seems to be to keep a bucket of each kind of bait and use one or the other or the two mixed, as the situation requires.

Many kinds of vegetables and fruit are used in making vegetable based baits, including ground and steamed taro and pumpkin. Soft coconut and avocado pears are also used and so is bread at times. A main requirement with any bait is to ensure that it is ground into fine particles, not left in small lumps which the fish quickly swallow, and which enables them to satisfy their hunger and then swim away. (See also 5.3.5.1 and 5.3.5.2.)

Fig 70   Underwater fishing lamp of Italian design with normal incandescent lamp bulb. The lead weight at the lower end of the protecting cage ensures that the lamp remains vertical under water

CHAPTER 6

HANDLING AND TRANSPORTATION OF LIVE BAIT

6.1 General

The survival of any species of live bait depends on a number of factors, the chief of which are:

— The external conditions of capture; *eg* day or night; when, where and how they are captured; water temperature, turbidity, bottom character, *etc.*
— The fish species and condition; *eg* their age and size, the degree of their gonad development, fat content, *etc.*
— The gear and methods used in capturing them.
— How they are handled on capture.
— The conditions under which they are kept afterwards.

6.2 Initial handling

One particularly important point is initial handling of the fish, when taken from the net and put into a holding cage or a tank. The fish are in a state of great agitation and can easily damage themselves if the conditions in the net are too crowded. The best action to take at this point is to allow the live bait a little while to calm down—say, 5 or 10 minutes or more—by leaving them in the net in which they have been caught before transferring them to a cage or a tank. Ideally, they should be able to swim into the cage from the net without any shaking of the net or touching the fish with a scoop.

If this type of transfer cannot be effected, as in the case when the fish must be put into a tank on board the fishing vessel, then buckets rather than brailers or scoops should be used. It is especially desirable where the live bait is delicate, such as in the tropics, and is subject to a high mortality rate because of abrasion and bruising, that the transfer is made by seamless buckets (15 to 18 litres) preferably made of stainless steel. The fish are kept in sufficient water in the bucket to reduce scale loss and bruising. Injury and shock during transfer account for most of the mortalities, and losses can be very high if the bait is handled carelessly. The use of buckets reduces losses.

When it is difficult to remove elusive fish with a bucket, the following method can be used: first, the fish can be taken with a scoopnet, but without removing the net from the water. Next, a bucket can be put under the scoopnet so as to contain it and the fish. Finally, the bucket with the net and the fish in water can be lifted and transferred to the receiver. In cases where fish pumps are used in this operation care must be taken not to allow the pump intake to get too close to the bottom and suck up mud. If this happens the mud will cause a high rate of mortality.

6.3   Floating and towed receivers

These are used for the transportation of live bait from the fishing area or other supply centres, (*eg* floating preservation cages, ponds, *etc*) to the tuna fishing vessels. Their use also allows the injured and weak fish to die before being transferred to the tanks on board.

The cage used by American tuna clippers for containing the live bait and towing it from the place of capture to the fishing vessel (*Fig 71*) has a box stern and a midship section of about 3m (10ft) in length, 2.3m (7½ft) in width and 0.9m (3ft) in depth, with a pointed bow section of about 1.8m (6ft) length. The bow is solidly planked, but the after section is made of netting on the bottom and sides so that the water circulates freely through it.

A similar but smaller type of wooden cage was used by FAO in the Caribbean. Its overall length was 1.83m (6ft) with a holding space of 1.52 × 1.2 × 0.71m (5 × 4 × 2.3ft) taking about 20 buckets of live bait. A large diameter bamboo pole was lashed to each of the upper sides to give buoyancy. The sharp bow section was made of plywood in which an eyebolt was fastened for the tow line. The sides and bottom of the cage were screened with minnow netting to permit water circulation and the container was towed at a speed of 2 to 3 knots.

There is also the *kowari*, a floating cage made with a wooden frame of four lengths of timber, 3.6 × 0.1 × 0.1m (11.8ft × 4in × 4in). The ends are

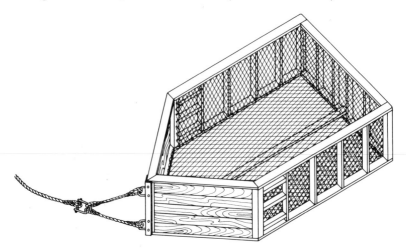

Fig 71   A towed live-bait receiver used by American clippers

notched and holed so that they can interlock and they are then held firmly with tempered wooden pegs. The cage can easily be assembled and just as easily dismantled for storing on board the fishing vessel. The knotless nylon net has four sides of $3 \times 3 \times 2.44$m (9ft 10in × 9ft 10in × 8ft). The net floats on its frame, the bottom corners being weighted with rocks to hold its shape and form a deep floating tank. A similar cage, but with bamboo frame (*Fig 72*) is employed by Japanese live-bait fishermen who use several types of towed cages.

A common practice is to transfer live bait to the cages and tow these to

Fig 72   A square, bamboo-framed floating live-bait cage used in Japan

Fig 73   Towing floating live-bait cages in Japan

a sheltered place where the water is unpolluted and anchor or moor them there. The towing is a carefully controlled operation, using a 200 to 240m towline and moving at a speed of not more than one knot, (*Fig 73*). This is to avoid excessive pressure on the cages, especially those made of netting, and to prevent the submerging of the upper frame. The towing is done by tugs powered by 100 to 200hp engines.

The larger the cages, the better the survival rate of the live bait. However, there is a limit to the size of cage that can be conveniently towed and handled.

Another type of net suitable for towing live bait is the towed container net shown in *Fig 74*. A much larger variant of this net has been developed

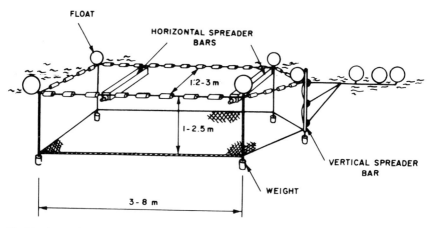

Fig 74   A towed container net

Fig 75   A square, floating cage from Sri Lanka

Fig 76   A floating live-bait basket from Sri Lanka

in Norway for towing fish into sheltered waters and holding them there until they have absorbed the food in their intestines and are ready for salting and canning. Clearly, this is a kind of net which could be used also in live-bait fishing.

Another type of square cage spread on crossed beams is used in Sri Lanka (*Fig 75*), where fishermen also use a basket tied to the side of their canoe or boat (*Fig 76*) as a bait receiver. The basket is made of loosely woven cane and is kept three-quarters submerged. Its main use, however, is as a live-bait tank during tuna capture operations. When the fishermen are searching for the tuna the speed of the boat is controlled so that sufficient water flows through the bait basket attached to the side of the vessel. Similar baskets are used also in Madeira. *Figs 77* and *78* show cane baskets used as bait holders in Lakshadweep.

In Japan, bamboo baskets are also used as fish receivers. The two sizes used in Kyushu, for instance, are 3 × 2.7 × 2.2m and 3.6 × 2.7 × 2.2m. More recently, baskets made of plastic have been introduced, but as they are more costly they have not been generally accepted.

Lakshadweep fishermen sometimes store live bait in a floating aluminium tank, 1.8 × 1.0 × 1.0m, such as in *Fig 77*. The sides are perforated to allow the water to circulate and two wooden poles are used to help float the tank when it is towed to the vessel or anchored in a lagoon.

In Hawaii, the skiff used for catching live bait has a bait compartment at its centre. The bait is brailed into this compartment and taken to the tuna fishing boat where it is transferred to the bait wells. If the skiff catches the

A. CANE BASKET

B. BAIT SCOOP

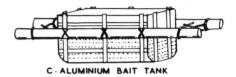

C. ALUMINIUM BAIT TANK

Fig 77 Live-bait handling gear in the Lakshadweep

Fig 78 A floating barrow-like live-bait basket from Lakshadweep

bait near to the fishing boat—which sometimes happens—the bait is towed in the net and brailed from there into the bait wells. A live-bait skiff propelled by a sculling oar is shown in *Fig 79*.

6.4 Preservation of live bait in floating cages

Once the fish have been captured the main problem is to preserve them until they are needed as bait in the tuna fishing. Such preservation other than on board the fishing vessels requires careful selection of sites in sheltered water (*Figs 80* and *81*). This is especially important where a specialized live-bait fishery exists, as for example in Japan, where fishing and preservation of live bait represent separate commercial operations.

When selecting a site, care must be taken to ensure that fluctuations in

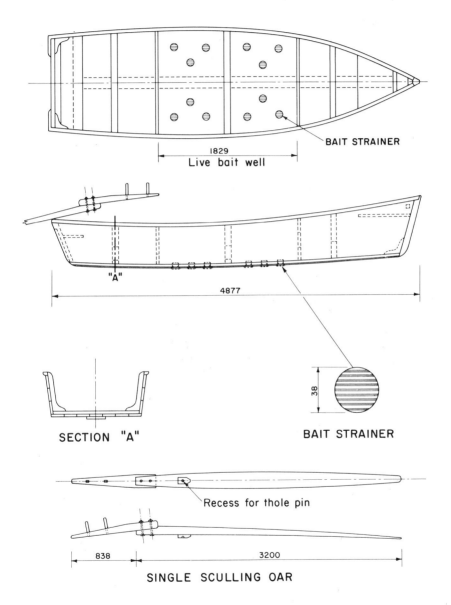

Fig 79 Skiff with well for live bait, propelled by a single sculling oar

Fig 80   Octagonal floating live-bait pens or cages in a live-bait preservation site in a protected bay (Japan)

water quality and temperature are as small as possible. It is also desirable that these characteristics should be similar to those of the water in which the fish were caught and that there should be a current to ensure a constant flow of fresh sea water through the cages; but not so fast as to shift them from their anchorages. The mooring site should be so well sheltered that the tuna fishing boats are able to come in and load the bait at any time. There should also be good land communication facilities.

The long experience of Japanese fishermen shows that for the best survival on board, the bait must live in the floating cages long enough to become acclimatized. The period needed for this is dependent on how long the fishing vessel must hold the live bait before use. For example, if the tuna fishing grounds are nearby and the boat spends only a day on the whole operation, the live-bait fish do not need to be in the cages for more than a couple of days. For long trips, however, they must be held in the cages for at least a week, preferably more. Indeed, in Japan, the live bait are rested in the cages for several weeks when they are to be used by tuna vessels that fish in the tropical western Pacific. During any rest period the weak and sickly fish will die, but the stronger ones that remain will become acclimatized and will survive their transfer and containment on board the vessel.

In some cases the fish are kept in the cages in which they are towed to the mooring place. The cages are moored in rows, tied to each other by steel and wire ropes, the first and last being anchored. If the first cage is located near to shore, it may be secured to a bollard or tree, or any other convenient mooring, in which case only the last outer cage is anchored.

A live-bait holder may be called a pen or cage (*Fig 80*) or by some other name, and it is usually made of a floating frame with a mesh box, cage or net bag hanging from it. The holders differ considerably in size and in some details of construction from place to place. For instance, a Japanese bait cage constructed of bamboo is usually rectangular in shape but is polygonal (multi-sided) when made of timber, usually cypress or cedar. The rectangular type (*Figs 73* and *81*) may be from 6 to 11m on the long side of the cage while the multi-sided variety (*Fig 80*) may be from 4 to 7m on each side. The bags may be made of 8 to 10mm minnow mesh or the normal 12 to 16mm stretch mesh netting. It may also be of 20mm stretch mesh for bigger fish, the choice being determined by the species to be caught, *ie* anchovy, mackerel or horse mackerel. The size of the bags differs, the square and rectangular bags being 3.6m deep, the octagonal 5.7m and the decagonal (10-sided) types being 7.2m. The minnow netting is usually made of cremona (PVA) or hemp palm fibre while the regular netting is made of cremona.

A problem associated with the small mesh netting is its tendency to clog, because various marine organisms adhere to it, especially algae. Such clogging slows the flow of water through the cage, which may lead to increased mortality. The best way to cope with this problem is to replace the netting at least once a month. The used netting can be cleaned, dried

Fig 81   Rectangular floating live-bait pens or cages in a live-bait preservation site in a protected bay (Japan)

Fig 82   Live bait, schooling in a floating cage

and used again. Care must be taken not to damage or frighten the fish in any way during this operation. The replacement of the netting once a month is suggested for normal conditions, but depending on the rate of clogging, it may have to be more frequent in many places while in others the replacement may be made at longer periods.

There are several ways of transferring the fish from one cage to another without harming them. One way is to place an empty cage alongside the one with the clogged netting, with the adjacent upper netting edges detached from the frames and either joined to each other, or with the clogged one overlaying the clean one. Both are then submerged and the fish swim into the new cage as the clogged one is slowly hauled up. A similar method can be applied with two netting bags, the clogged and the clean one, hung from a single frame.

Bird predators can be a nuisance and a source of loss, and bait fish can be protected against them by covering the containers with netting. Where such covers must float, they should be made from buoyant material such as polyethylene or polypropylene. Another problem can be the intrusion of predatory fish, which have to be removed one by one, as no means has yet been found for easier separation of live bait from their enemies. There is also a need to keep some personnel at such a site, working with one or two small boats to replace clogged nets, feed the fish and catch predatory fish. They would also transfer the live bait to the fishing boats, maintain mooring systems, communications and so on.

Reference has already been made to the fact that transfer of the live bait to the floating cage, and towing it to the fishing boat or mooring place greatly agitates the fish. The smaller the fish and the higher the temperature during this operation, the more the fish are at risk and as with all live bait, the longer the period of towing the higher the mortality rate. Japanese statistics on live-bait mortality show that 50 to 70 percent of anchovy juveniles up to six months of age die in this initial holding period, whereas the mortality rate for two-year-olds is between 30 and 50 percent. Anchovy caught immediately after spawning suffer a very high mortality, but in winter the mortality rate for all ages drops to less than 30 percent.

6.4.1 FEEDING LIVE BAIT IN CAGES

Correct feeding of live bait while they are held in cages pending transfer to the fishing boat and, later, while kept on board, is essential for keeping down the mortality rate. The fish are not likely to want to feed for some time after capture—probably not until three or four days in captivity if they are small (less than 5g weight), five to six days if they are of medium size (5 to 6g weight), and seven days or more if they are bigger. There is no difficulty in finding out when they should be fed because they will indicate this by schooling and swimming around in circles (*Fig 82*), and by excited response to feeding (*Fig 83*).

Fig 83   An excited response to feeding

There is a problem, however, in providing the correct daily ration. This varies in quantity according to the size and body weight of the fish, the type of feed provided, the response of the fish to the feed and the temperature of the water. It is estimated that the daily ration of compound food, given in two or three feedings a day should be about 2 to 3 percent of the body weight of the fish, and about 5 percent of the weight when the feed is a natural substance. When the water temperature increases and the fish are responding well to the feed, the ration should be slightly increased. When the temperature decreases and the response of the fish to the food declines, the ration should be slightly decreased. It may also be a good thing to vary the diet, changing from one type of food to the other as well as trying out various mixes of them.

Compound feed usually consists of more than 50 percent crude protein—mainly fish meal—and sugars, crude fats, crude fibres, ashes, water, *etc.* The natural feed generally includes frozen mysids (*Mysidacea*) and frozen fish such as chopped or minced mackerel, anchovy, *etc.*

It is recommended that compound feeds should normally be used because they are easily stored, can be medicated without difficulty if so required, and are easy to dispense. The feed should be put directly under the water, not sprinkled on the surface, because the falling particles excite the fish which, in their agitated movements, could suffer body injury. However, care must be taken to avoid giving too much feed because that which is left uneaten may contaminate the water, causing disease and increased mortality.

6.5   Live-bait tanks on board tuna fishing vessels

The design, construction and size of live-bait tanks or wells is determined by the size of the boat, its fishing range, and the technological level of the fishery. They can be roughly classified into three main types; bilges or compartments in the boat used as live-bait wells (*Figs 19* and *21*); a well specially designed to form an integral part of the boat's structure (*Fig 26*), and tanks installed under or on the vessel's deck (*Figs 15* and *16*). The first two types are used in sailing boats and in small to medium size motorized vessels, but not in the big ocean-going vessels. The latter mostly have more complex tank systems linked with a refrigeration system.

6.5.1   KEEPING LIVE BAIT IN BILGE WELLS

The Indian Ocean is one area where fishermen use the bilges of their boats, sometimes divided into compartments, for bait carrying. An example of this type of bait holding is to be found in the traditional boats of Lakshadweep, where the deep bilge in the middle of the boats is divided into four or five compartments for storing live bait. The water enters the middle compartments through holes bored in the hull and flows into the adjoining compartments through perforations in the partitions, thus maintaining water circulation. The live bait, which are removed by net when wanted, are kept in the middle compartment, the water being bailed out from the end compartments.

The pole-and-line fishermen of the Maldives have a similar arrangement for keeping live bait in their boats (*Figs 21* and *22*).

6.5.2 BUILT-IN LIVE-BAIT TANKS

In the Japanese small-scale tuna pole-and-line boats, the built-in live-bait tanks vary in capacity according to the size of the boat. For example, a 5t boat has three tanks of 1cu m, while a 10t vessel has three tanks of 2 to 3cu m each. There are four to six holes in the bottom of each tank through which sea water circulates as the boat moves through the water. When the sea is so calm that water circulation is not sufficient, flat pieces of wood or metal, about 20 to 30cm in length, are fixed at the holes so as to protrude into the sea, thus increasing the amount of water entering in the holes. The tanks are used as fish holds as well as for live bait. The holes in the bottom of the tanks are of two kinds—round in the bigger tanks, and rectangular in others. When a tank is used for live bait, a screen is placed over each hole to prevent the fish escaping. When a tank is used as a fish hold, the holes are plugged up. Another example is the live-bait well in pole-and-line boats designed by FAO (*Figs 26* and *27*), where the insulated well is an integral part of the boat.

Traditional arrangements of keeping live bait in bilges are gradually being replaced by live-bait tanks as motorized fishing boats are introduced. In Lakshadweep, motorized vessels are fitted with a bait tank of 1.6 × 0.8 × 0.8m, which is fitted forward of the engine room. The tank is partitioned in the middle by planks drilled with holes, an arrangement which makes the catching of the live bait easier when they are required for the chumming operation. Water is circulated by inlet and outlet tubes and a detachable metal strainer is fitted on the inner side of the outlet pipe to prevent the escape of fish. When the boat is under way, the water is pushed into the tank through the inlet pipe, but when the boat is at anchor or not making way, the water has to be pumped in or poured in by hand. The stopcock is closed at such times—or the rubber inlet tube is tied—to prevent the tank from draining. A tank of this size can be both an integral part of the boat or a separately installed container.

6.5.3 LIVE-BAIT TANKS FOR VESSELS EQUIPPED WITH ENGINE-DRIVEN WATER PUMPS

These tanks are suitable for smaller vessels being built or converted for pole-and-line fishing and the size of the tank or tanks is limited by the size of the vessel. When installing tanks, especially on deck, great care should be taken not to raise the centre of gravity of the vessel to such an extent that it becomes unstable and therefore dangerous.

So far as possible, the tanks should be positioned to give the vessel the most suitable trim for fishing. From the point of view of stability, the tank should be placed so that as much of it as possible is below deck level, the ideal being to have only the hatch coaming projecting above deck.

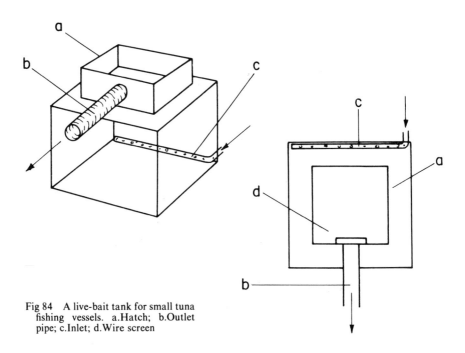

Fig 84   A live-bait tank for small tuna
fishing vessels. a.Hatch; b.Outlet
pipe; c.Inlet; d.Wire screen

6.5.3.1   *Construction*

The tank and hatch coaming may be constructed of steel, timber or heavy waterproof plywood (*Fig 84*). The hatch coaming should be of sufficient height so that there is always some water in it even when the vessel is rolling heavily. If, under rough weather conditions, the water leaves the hatch, the excessive swirl and slapping of the water under the shelf of the tank will injure the bait and increase mortality. Maintaining the water in the hatch at all times is also an important factor in the stability of the vessel.

The inlet for the water supply consists of a perforated pipe blocked off at one end. This pipe can be laid along the bottom edge of the tank opposite the outlet, with the holes directing the flow of water across the floor of the tank, (*Fig 84*). However, in some cases it has been placed along the opposite or adjacent sides of the tank, or standing upright in a corner with the flow of water directed along a side wall. The holes in the inlet pipe should be big enough to allow the water to flow in evenly, without causing excessive turbulence.

The outlet or overflow pipe is situated well up on the outboard side of the hatch coaming and will need to be at least twice the diameter of the inlet pipe. It is backed, on the inside, by a large screen of 6mm square galvanized wire mesh raised 25 to 50mm from the inside face of the coaming, (*Fig 84*).

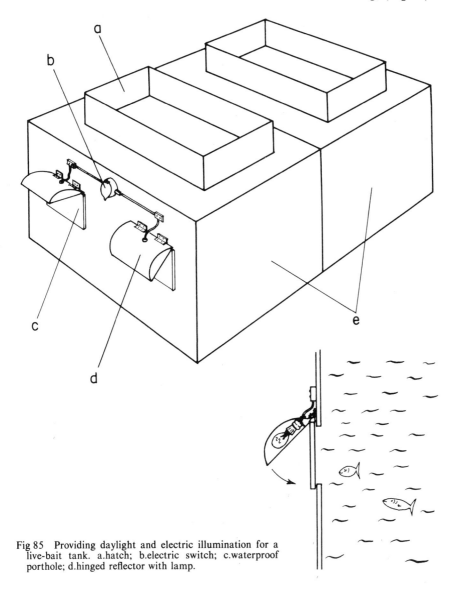

Fig 85   Providing daylight and electric illumination for a live-bait tank. a.hatch; b.electric switch; c.waterproof porthole; d.hinged reflector with lamp.

6.5.3.2 *Illumination*

In (6.6.3) the importance of a correct level of illumination in a bait tank for the survival of the fish is more fully discussed.

The light should preferably be hung above the tank at dusk and can be a type of pressure lantern or an electric lamp. If possible, submerged electric lamps should be used, one or two of them installed at the tank's sides or under the shelf. Where tanks are installed on the deck, the light could be provided from outside, by providing one or two waterproof portholes halfway down the side of the tank. These would provide light during the day, while a lamp in a hinged reflector (*Fig 85*) on the outside can then be used to throw a beam of light directly into the tank during the dark hours.

6.5.3.3 *Pump system*

The pump should be placed below the water line of the vessel so that it is self priming. The sea inlet should be well down towards the keel and within the midship section of the vessel. It is important that no air at all is sucked through the pump. Air going through the pump is broken up into very small bubbles, which give the water a milky look, and kills the bait fish in a comparatively short time.

The capacity of the pump must be sufficient to change the seawater in the tank every six to eight minutes. For example, a $1.20 \times 1.20 \times 1.2$m tank suitable for a small to medium size fishing boat has a capacity of approximately 1.7cu m. This quantity of water must be changed up to ten times an hour; therefore the pump must have a capacity of 17cu m/hr.

A simple pump system consists of sea inlet, sea cock, pump, delivery cock and the delivery line to the tank. Its disadvantage, especially if the tanks are set low in the hull, is that when draining the tanks only some of the water will run back through the pump. Most of it must be bailed out or else drained into the bilge.

Larger boats can use a more suitable system (*Fig 86*). For normal operation the sea cock and delivery cocks are opened and the by-pass and discharge cocks are closed. In order to pump the tanks out the discharge, by-pass and tank delivery cocks are opened and the sea cock and main delivery cock are closed.

With this system it is possible to fit an emergency bilge-draining line. To pump out the bilge the delivery cocks and the sea cock are closed and the bilge, by-pass and discharge cocks are opened. When pumping the bilge it may be necessary to prime the pump by opening the sea cock or, if there is water in the tanks, the tank delivery cocks. They can be shut again as soon as the pump starts working.

As a safety precaution, if an emergency bilge suction line is fitted the cock should be wired shut when not in use. It will be seen from *Fig 86* that with the sea cock, bilge suction cock, and either the main delivery or by-pass cock open, the sea water can flow directly into the bilge.

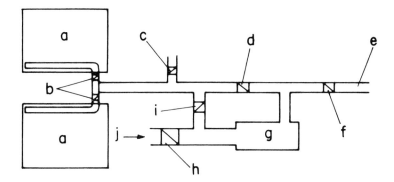

Fig 86   A live-bait tank for a larger tuna fishing vessel
a.Bait tank; b.Tank delivery cocks; c.Bilge suction cock; d.Delivery cock; e.Discharge overboard; f.Discharge cock; g.Pump; h.Sea cock; i.By-pass cock; j.Sea inlet

6.6   Preservation of live bait in tanks on board

A number of factors affect the survival of live bait in tanks on board. These include oxygen content in the water and oxygen consumption, lighting, water temperature and its changes, and other qualities of the water such as pH *etc.*

6.6.1   TRANSFER AND INITIAL STAGE

The first consideration, of course, is the safe transfer of the live bait to the tanks (*Fig 87*) when, as previously described for the transfer of the fish from the net to a floating container, buckets should be used (*Fig 88*).

The live-bait fish become agitated during transfer, resulting in an increase in the consumption of oxygen, in excretion and in loss of scales, a situation which could increase pollution in the tank. A careful watch over the fish in the ensuing period is required so that the necessary action can be taken to control conditions and keep pollution to a minimum. A Japanese study has indicated that more than 70 percent of live bait die if the state of agitation continues for 50 hours.

A Hawaiian study on factors affecting the mortality of Hawaiian anchovy *Stolephorus purpureus* has shown that the initial mortality from injury, shock, and stress is complete by the first day and survivors were preserved for as long as 23 days, under controlled conditions.

The early American tuna-clipper fishermen might well have been aware of this condition for, after catching bait, the clippers usually remained for a day or two in quiet water for a 'rest period' to enable the fish to settle down to life in the tanks.

A secondary or delayed mortality may take place anywhere from 2 to 14 days after capture, but this may be prevented for long periods by controlled environmental conditions.

The environmental conditions can be made more favourable by increasing the amount of oxygen in the holding tanks, reducing the temperature and

Fig 87   Transfer of live bait from a floating pen in Japan
a. Setting the liftnet; b. Hauling and concentrating the fish for brailing

Fig 88   Transfer of live bait to tank on board

lowering the salinity, at the same time avoiding overcrowding the fish and avoiding excitement, in order to help them settle down.

The Hawaiian study recommends the following:

Salinity: 50 percent sea water (approx 16.0–18.0 ppt) for as long as possible, especially during the first three hours after capture. Open system circulation preferable, but a closed system should recirculate brackish water for no more than three hours at bait density not to exceed approximately 220g/100 litres (1.6–1.8lb/100gal); pH should not fall below 7.5.

Oxygen: supplementary oxygen desirable; concentration not to exceed 10ppm or to fall below 5ppm. Ideally, oxygen should be released through porous stone at a pressure sufficient to release minimum-sized bubbles for optimum solubility in sea water.

Temperature: 21–24°C; higher temperatures to be avoided, if possible. Cooler water evidently slows metabolic processes and oxygen requirement, thereby easing stress and permitting bait to overcome initial shock more easily.

Current: horizontal current; approximately 0.10m/sec (15-20ft/min); not to exceed 0.13m/sec (25ft/min). Decrease current speed for smaller or injured fish to minimum required for orientation on first day. Maintain current at least until period of delayed mortality is over. Current speed in the tank never to exceed 0.3m/sec.

Light: white or green light; light at night important, *eg* subsurface 40W both night and day.

Green light appears better to promote normal orientation and milling, and to decrease stress and increase survival.

6.6.2   CIRCULATION IN THE TANK

It is easy to watch the fish, as the tank is lit by a 40 to 100W underwater lamp placed at the tank centre. The water and oxygen are circulated in the tank and should be adjusted according to the behaviour of the fish.

If air bubbles increase too much, the fish will descend in order to get under the bubbles, and their normal swimming pattern will be disturbed. This bubble condition must be corrected by regulation of the air flow. This is important because an excess of bubbles may damage the live bait by sticking in the gills and suffocating the fish. It may also cause collisions between the fish through upsetting their normal swimming pattern, and bubbles can become attached to their bodies and can adversely affect the functioning of body fluids.

The above-mentioned Hawaiian study also comments on the following:

Circulation and flow rate in an open system: maximum flow rate possible without exceeding optimal current speed at optimal turnover rate of 50–60 percent of the tank volume per hour and not less than 40 percent per hour.

Recirculation in closed system: 50 percent of sea water recirculated for not more than three hours at a bait density of not more than 180g/100 litres (1.6–1.8lb/100gal) unless a filtration system is added to remove ammonia, and a buffer used to sustain pH at about 8.0. Add protein skimmer to remove organic particles, mucus and scales. Obviously, open systems, that is where new water from the sea is continuously supplied, are preferable and more feasible on board fishing vessels.

As a general rule, American skipjack fishermen maintain a change of the water in the wells at five to ten times an hour. A full change of water every eight minutes is a widely accepted practice.

6.6.3   LIGHTING

Reference has already been made to the underwater light in the tank, proved necessary because the total mortality of the live bait exceeds 50 percent if the tank is kept dark all the time. The use of natural or artificial light reduces the mortality to less than 10 percent. However, the optimum illumination has not yet been established, which is all the more reason why the behaviour of the live bait should be watched to try to establish the best level of light. This may call for some experimental use of lamps, such as the use of two 40W lamps in a 20cu m tank or using lamps of different colours. It was found that Japanese anchovy became quieter with a red lamp than under a white light.

6.6.4   DENSITY IN THE TANK

The quality of live bait carried by a fishing vessel must be determined by several factors such as the size of the vessel and live-bait tanks, the duration and intensity of the fishing, and the prevailing temperatures. For example, a pole-and-line boat that is going to fish skipjack for a long period in warm waters will take less bait per tank volume than a boat that is going to fish for albacore in cold water grounds. According to the Japanese practice a recommended tank density of fish for fishing in warm waters is 10kg/cu m. That for fishing in near or offshore waters is 15 to 20kg/cu m for trips lasting about a week. The density recommended for albacore fishing in water temperatures of, say, 19 to 22°C, is around 20kg/cu m.

The density recommended by the Hawaiian study for the more delicate Hawaiian anchovy is approximately 2.0kg/cu m although at a rather slow water turnover rate of 50–60 percent per hour.

6.6.5   TEMPERATURE

One important cause of increased mortality is any major or rapid change in temperature. As has already been stressed, the longer the fish have to adjust to such changes, the less the mortality. This suggests that when the vessel sails into an area with different water temperature the circulation in the tank should be slowed down, thus limiting the in-flow of seawater and steadying the rate of temperature change. Another suggestion is to put fewer fish in a tank, so that more oxygen is being made available to each fish, a practice which could also be combined with the slowing down of water circulation.

Some conditions which adversely affect the live bait are, to a large extent, unavoidable. A case in point is a voyage to warmer waters when the sea temperature rises 3 to 4°C/day. A study on such a voyage showed that in 3 days after departure, during which the water temperature in the tank rose by 7°C (from 21.1°C to 28.2°C), the body fat and water content of the fish decreased rapidly, as did the oxygen content, and 50 percent of them died.

6.6.6   WATER QUALITY

The quality of water in a live-bait tank should be controlled as carefully as possible. A study was made of dissolved oxygen (DO), pH, biological oxygen demand (BOD) and NH_4 in a tank during a voyage. It was noted that the DO was lowest immediately after the fish were put into the tank but gradually recovered while BOD changed at once and reached the worst condition in 30 minutes. Then it recovered but reached a high value, which is not a favourable condition. The level of NH_4 was highest immediately after the fish were put in but gradually decreased as a result of water exchange. The rapid increase in excretion due to the sudden change in the environment appears to be the cause of this. As a consequence, CO_2

increases and pH value drops from 8.3 to 7.3, at which the fish can survive
if the decrease is gradual.

Filtering the water is an essential for quality control, along with the
required adjustment in circulation flow to meet the demands of the changing
conditions. All this, however, depends on the species; according to the
Hawaiian study a pH of 7.5 is lethal for the Hawaiian anchovy. It should,
therefore, be maintained at 8.0–8.1 level and never allowed to fall below
7.6.

6.6.7   FEEDING ON BOARD

The live bait, of course, also need to be fed on the voyage to the fishing
grounds if the trip takes more than a few hours. It is recommended that
they should be given three feeds a day, the amount of feed being determined
by the size of the fish and the water temperature. For example, the amounts
of food per day by percent of body weight for Japanese anchovy at different
water temperatures have been established at: below 18°C, 1 to 2 percent;
18°C to 24°C, 2 percent; 24°C to 28°C, 2 to 3 percent; over 28°C, 3 to 4
percent.

6.6.8   FUTURE IMPROVEMENTS

It can be said that much work remains to be done in finding ways and
means to bring down the usual rates of mortality among live bait from the
time of their capture to their use. There is need to find better means of
transferring the fish, such as through the use of pumps, and to determine if
drugs can be used to reduce the fright reaction and metabolic requirements
of the fish, just as anaesthetics and tranquilizers are used in transporting
trout, salmon, *etc.* Meanwhile, the practices recommended here at least
provide proved ways of handling and preserving live bait.

CHAPTER 7

TUNA FISHING GEAR AND AUXILIARY EQUIPMENT

Whatever the size of vessel engaged in pole-and-line fishing for tuna, the basic gear required is the same—hooks, jigs, lines and fishing poles. There are, of course, many variations in the gear, such as the angling machine, as well as in the range of auxiliary equipment.

7.1 Hooks and jigs

7.1.1 HOOKS, SINGLE

The hooks used in tuna pole-and-line fishing are not, as a rule, barbed (*Figs 89* and *91*). If they have a barb a small one is preferable, since it is much easier to unhook a fish from a barbless hook. As there is intense pressure on the fisherman to catch as many fish as possible while the boat remains with the school, quick unhooking is essential.

The Japanese hooks used for catching small to medium size fish measure from 3.3 to 3.6cm while those for the larger tuna measure from 3.5 to 6.4cm. There are some regional differences in their shapes (*Fig 89*) and western-type hooks generally have longer shanks and are mostly barbed (*Fig 90*).

As already noted, there are many variations in the hooks, for example *Fig 91* shows hooks used in Sri Lanka and in Lakshadweep. The hook favoured by Polynesian fishermen in the Pacific Islands for fabricating the pearl-shell lure is made from 3mm diameter high tensile stainless steel wire and is usually 40mm longer than the shell part, the tip being almost in line with the front of the lure (*Fig 92*). In many places the hooks themselves are still made from material such as shell.

7.1.2 HOOKS, DOUBLE

As a rule, forked double hooks are used for trolling, their size varying with the size of fish; largest for the yellowfin and albacore tuna, smaller for the skipjack, and the smallest for the frigate mackerel. Both barbed and non-barbed hooks are used (*Figs 93* and *94*).

7.1.3 JIGS OR LURES

The jigs for both pole-and-line fishing and trolling are designed with special attention to their shape and colour as they are mainly intended to appeal visually to the fish. Nevertheless, it is not always necessary for the jigs to exactly simulate the appearance of tuna prey. Instead, a substance looking roughly like a fish is enough for the purpose, so far as the shape is concerned (*Fig 94*). On the other hand, a suitable arrangement of colours, much brighter than the actual ones, is needed, because colours appear to be more important than shape in enticing the fish to strike. In addition, the effect of a jig will be enhanced by quick and continuous movement in and out of the water, as is the practice in tuna angling. It also explains why fish appear to strike better in a rippling sea than when it is calm. While trolling, an elongated small float is sometimes inserted between the main line and the snood to encourage a 'porpoising' or jerking motion of the jig.

When fish are as excited as is usually the case in pole-and-lining for skipjack and frigate mackerel, jigs are preferred to bait and hooks, in order to save time and labour. Even where jigs are not as effective as real bait, fishermen keep them handy for use when the stock of live bait on board is exhausted, especially in trolling for tuna or yellowtail.

Most jigs are provided with a shaft and a single or a double hook. Exceptions are those which have only a polished hook or a hook furnished with fish skin or a glass bead.

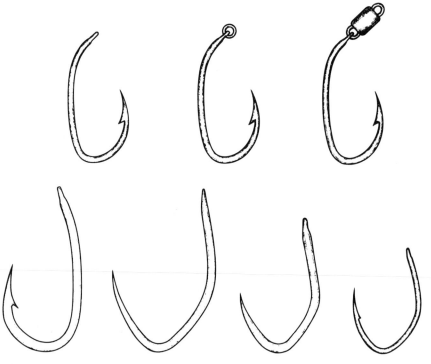

Fig 89   Japanese type tuna hooks

The jigs, or various kinds of lure, are made from a great variety of materials, such as horn, hoof, bone, plastic, rubber, zinc or wood and painted with powder of mother-of-pearl. Bird or synthetic feathers, fish skin or seaweed are used for the tails of the jigs.

A rather new type of jig-tail is thrust over a hook in the same manner as real bait. With the recent application of plastics and sponge rubber to fishing gear, this type of jig is being produced to look like squid, crab, saury or fish eggs. Sometimes they are used together with, and sometimes independently of, true baits.

7.1.3.1 *Japanese feather jigs*

A jig popular in Japan is roundish, made of lead or brass and tin-plated, the thickness of the jig being in relation to the fish to be caught. The hook is mounted on a shaft and is covered with natural or nylon-imitation feathers, mostly white but with some red and brownish ones. These are bunched around the shaft. The feathers are covered with a piece of fish skin, such as that of dolphin and leatherfish or catskin.

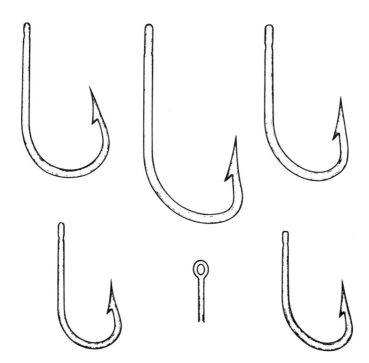

Fig 90   European type, barbed tuna hooks

7.1.3.2 *Pacific Island feather jigs*

In Fiji (*Fig 95*) fishermen use an artificial jig and a *Zundo* tuna hook embedded in lead in a brass tube, an eye being fixed to the tube. The hook and shank are chrome finished and white chicken feathers are secured to the shank of the hook, with cat or dolphin skin fastened to the indented part of the tube. The skin forms a wrapping around the base of the feathers to prevent fraying.

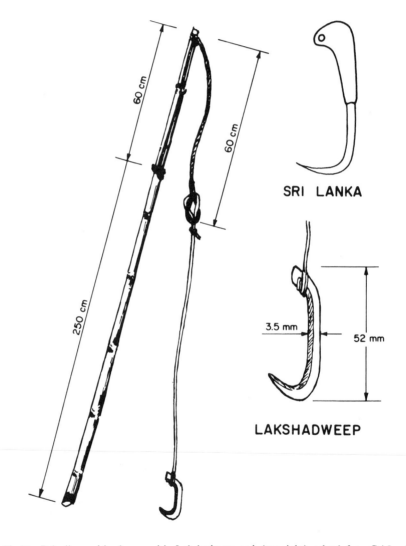

SRI LANKA

LAKSHADWEEP

Fig 91   Pole, line and hook as used in Lakshadweep and, (top right), a hook from Sri Lanka

Fig 92 Polynesian hook with pearl-shell lure

A similar arrangement is made by Hawaiian fishermen with various coloured feathers or threads being used for the lures.

7.1.3.3 *Pearl-shell lures*

In Tahiti, pearl-shell lures are used (*Fig 92*). They are made by hand, a job requiring a high degree of traditional skill. It is claimed that these lures are

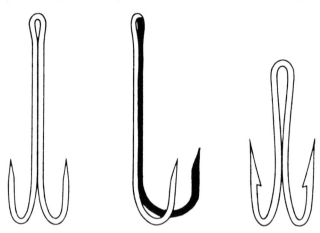

Fig 93   Double hooks

much more effective than any mass-produced variety. They are made from the shell of black-lip oysters (*Pinitada margaritifera*) but the Australian gold-lip pearl oyster (*P maxima*) can be used, being even more suitable because of its white and gold colour. It is said that not more than one in 200 Tahitian pearl shells, perhaps even less, is suitable for making lures, which indicates the care needed in selecting a shell, which must be at least 10 years old.

Four or five lures can be made from one valve, the shell being marked by the lure maker to show where it must be cut (*Fig 96*). The slivers are ground into shape, the top skin being ground off until a yellowish colour is reached. A hole of about 2mm diameter is drilled through the tip and a small notch made on each side near the end. Most lures are 10–12cm in length and 10–18mm in breadth, but the respective size ranges are 5–20mm and 8–22mm.

The boats always carry a wide selection of lures prepared on snoods, anything from 30 to 100, varying in colours and sizes. One kind, a small metal lure called a *champignon*, is used for small skipjack.

7.1.4   TROLLING JIGS

Most pole-and-line fishermen in various parts of the world also troll as a way to locate and attract tuna schools. Japanese fishermen use for this purpose jigs which in many ways are similar to those used with the poles (*Figs 97* and *98*), except for double hooks and a slightly bigger bunch of feathers, perhaps with small pieces of shell attached to the sides of the lure (*Fig 97* - top). Again, on occasion, artificial lures without hooks are trolled to attract tuna toward the fishing boat. Such lures, mostly 10 to 20 plastic (generally vinyl) squid or octopi, are tied to 3–5m nylon snoods which are

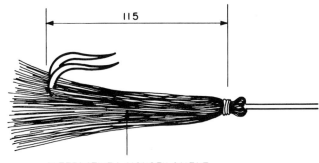

ALTERNATIVE : MONOFILAMENT
HORSEHAIR / MAIZE

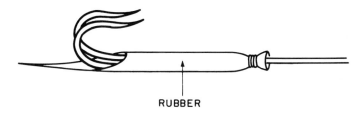

RUBBER

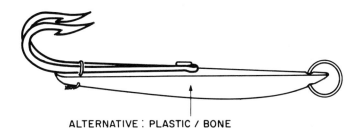

ALTERNATIVE : PLASTIC / BONE

Fig 94 Double-hooked tuna jigs. The upper two jigs are used for angling and the lower jig is for trolling

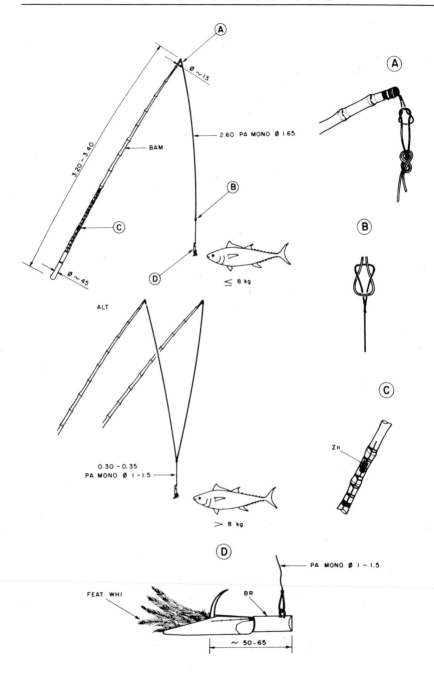

Fig 95   Tuna angling gear from Fiji

fastened to the trolling line. A large variety of trolling jigs is available commercially.

7.2   Lines

The lines used with the poles are mostly made of synthetic fibres, although in some areas lines made of local natural materials are still used. Such traditional lines may have a leader made of hemp and a snood made of gut or wire. Nowadays, however, most lines are made of nylon monofilament (PA mono). The lines, (leader and snood), are usually about 40cm shorter than the poles, though even shorter lines are sometimes used, as these enable the fishermen to swing the fish more easily overhead and unhook it by slackening the line (*Figs 91, 95, 96*).

In some areas, pole-and-line fishermen used to have a swivel with a quick release device spliced into the lower end of the leader. In such a case a snood of stainless steel wire, made up in two pieces, slips on to the quick release. The top piece of the snood should be from 38 to 50cm long and the lower piece from 10 to 15cm. The jig, or a live-bait hook, is attached to the lower piece of the snood.

If all snoods are of the same suitably adjusted length, it is not necessary to alter this each time a new jig is required or a change is made to a baited hook. A suitable length of leader will allow the jig to reach to a point approximately 40–45cm short of the butt-end of the pole.

In Lakshadweep, a R4000 tex (approx) cotton twine is used for the leader with 1mm diameter nylon monofilament for the snood (*Fig 91*). In Fiji, the leader is 2.6m in length, made of 1.5–1.7mm diameter monofilament nylon, with a 30 to 35cm (1.0 to 1.5mm diameter) snood. The leader is attached to the main line by looping.

Madeira fishermen have been using 1 to 1.5m brass wire-rope snoods, made up of seven to ten wires. These are tied to 2m leaders for tuna and 1m leaders for skipjack fishing (*Fig 99*).

The tackle used by Polynesian fishermen using pearl-shell lures (*Figs 92 and 96*) includes a 2.90–5.00m line made either of 3 to 4mm diameter cotton, or of 1.5mm diameter nylon monofilament with a snood 80 to 100cm in length, ending in a loop for easy attachment to or detachment from fishing lines. The snood is made from 2.5mm diameter braided nylon twine or 1.5mm nylon monofilament, and is bound firmly along the roughened part of the hook shank with nylon twine or waxed cotton. The twine is then passed several times through the hole drilled in the lure tip and around the shank of the hook before being neatly tied. The rear of the hook is then bound firmly to the lure, the twine being slotted into the small notches made in the lure near the end.

Trolling lines used during pole-and-line fishing operations may consist of 3 to 5mm diameter mainline and 0.5 to 1.0mm diameter monofilament nylon leader with a 0.4 to 0.7mm diameter filament nylon snood. The length of the mainline depends upon the type of operation and the length of the

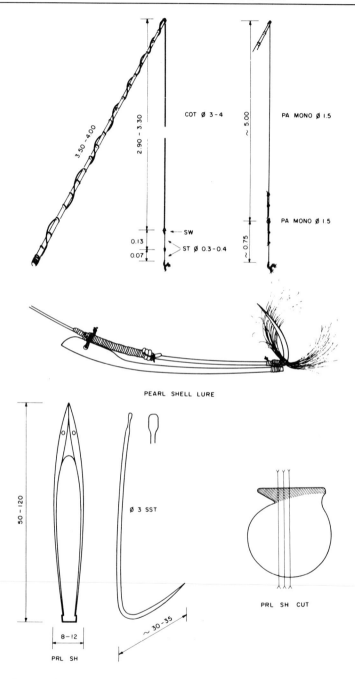

Fig 96   Tuna angling gear from Tahiti, showing details of pearl-shell lure

vessel. It should be 20 to 50m long when used during the passage to and from the fishing ground, but while on the fishing grounds, especially if the schools are dense and reacting well to bait, the mainlines can be shortened down to 10 to 12m. Leaders are up to 15m long and the snoods are 1 to 2m long. Swivels may be inserted between the leaders and the mainline and

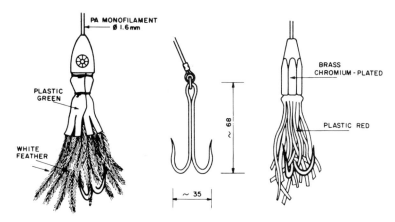

Fig 97   Trolling jigs from Japan

between the leader and the snoods. Shock-absorbing cords made of sections of old automobile tubes, 30 to 70cm long, are inserted parallel to the mainline, where it is attached to the vessel or to a special outrigger. Wire snoods, preferably of steel piano wire, are sometimes used.

7.3   Poles

Typical smaller boat bamboo poles are from 2.40m to 3.00m long, approximately 50mm diameter at the butt, and tapered toward the tip. They are seized and bound with adhesive tape between the knots to prevent splitting. One way of connecting the leader to the pole is to have a bight of

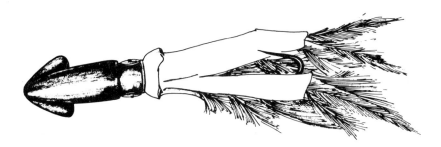

Fig 98   A squid-shaped jig from Japan

heavy, hard-laid cord seized to the tip of the pole. The leader is attached to this bight with a double sheet bend, and any adjustment to the length of the leader is made at this point.

In the past, only bamboo was used for making the poles in Japan but today most of them are made of GRP or FRP,(fibre-reinforced plastic or fibreglass, *Fig 100*). They range from 2.50 to 5.65m in length, depending on the weight of the fish and the nature of the fishing, whether baited or baitless jigging, or for automatic fishing machines (*Fig 9*). There are also some regional differences in length and thickness.

The poles should be selected by their users, the length and weight being determined by their skill and strength as fishermen. The butt end of the pole is bound with thread to provide a firm gripping surface and prevent the

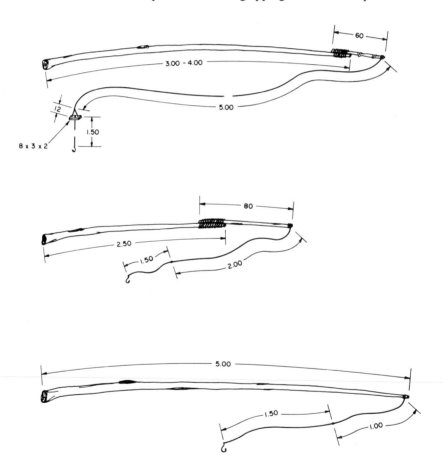

Fig 99 Angling gear from Madeira. Top, *vigia* gear; centre, tuna tackle; bottom, skipjack tackle

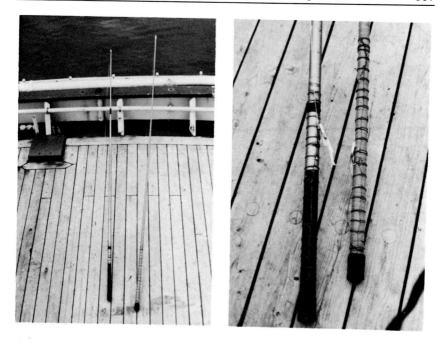

Fig 100 and Fig 101   Japanese angling poles made from glass reinforced plastic (GRP) and right, butt-end grips

Fig 102   Long pole in use off Tahiti

fisherman's hands from slipping, as they might do on the smooth surface of
the pole. The thread also holds the hook when it is not in use (*Fig 101*).

The fishermen of Madeira mainly use two types of poles. The tuna pole is
made of laurel tree wood, 2.5m in length, to which is fastened an 80cm long

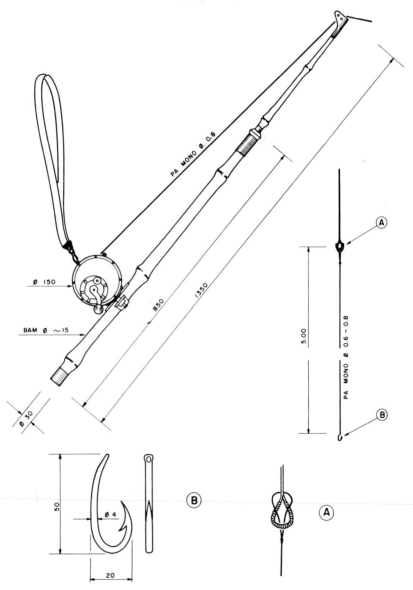

Fig 103   Sport type tuna tackle, from France

rod of quince tree wood. The skipjack pole is longer. A third type, called *vigia* is used rather for detection than for fishing the tuna (*Fig 99*).

The poles used in Hawaii vary from 2.3 to 4.6m in length, the shorter ones generally being used to catch fish of 9 to 10kg or more. In the Maldives, the poles are about 3.60m in length, while those in Fiji are usually about 3.20 to 3.40m tapering from about 45mm diameter in the butt to 13mm at the tip. Sri Lanka fishermen use poles of 3 to 4m in length with a mean diameter of 35mm. In Tahiti (*Figs 96* and *102*) the length of poles varies from 4 to 8m, the butts being about 80 to 90mm diameter tapering to around 20mm at the tip—perhaps a bit more, but not less. Many Tahitian boats carry 10 or more short poles, no more than 3m in length, used for fishing small skipjack with small metal lures known as *champignons*. Most of these poles are still made of bamboo.

French fishermen angling for tuna with live-bait attraction use sport type tackle (*Fig 103*) with a winding reel mounted on a rather short strong pole.

7.4 Auxiliary equipment

The auxiliary equipment carried by pole-and-line fishing vessels includes handnets, scoopnets, bait buckets and bait chumming tanks.

7.4.1 HANDNETS AND SCOOPNETS

The handnet or 'crowder' (*Fig 104*) is used for crowding live bait in the holding tanks into a small space so that it can readily be scooped with a scoopnet. It consists of a rectangular section of 12mm mesh netting, a bit longer than the width of a tank and about 1m deep. It has a double handle,

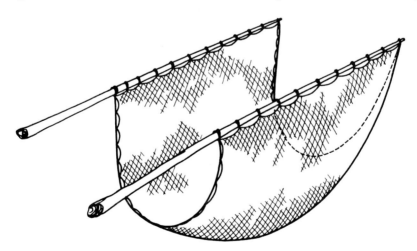

Fig 104   A 'crowder' handnet

that is, one pole on each side of the net, so that the net can be used with both hands. The double handle is about 1m longer than the net side, thus making it easy to handle.

To use the crowder, fishermen hold the poles apart and move them along in the tank until the desired number of fish swim into the bag of the net. Then they bring the poles together, lift them out, and drop one on each side of the hatch so that the bait is confined in the hatch.

Scoopnets are used for several purposes, depending mainly on their size. The large ones, which are about 40cm in diameter and 20 to 30cm in depth, are generally used to transfer the live bait from the tank or crowder into buckets or to catch non-bait fish in order to remove them from the live-bait tanks. They are also used instead of buckets in some places for transferring the fish from live-bait cages to the holding tanks on board. The small scoopnets (*Fig 105*), which are about 20cm in diameter, are used in chumming (*Figs 106* and *107*).

7.4.2 CHUM TANKS AND BUCKETS

Live-bait buckets are made of wood or a plastic material and some are 25 to 30cm diameter and only 20cm in height. They are used for keeping a few live bait (up to about 10) at the side of the fishermen for use in chumming during the fishing operation.

Live-bait chumming tanks are usually put on the deck near to the angling

Fig 105　Colander-shaped small scoopnets from the Maldives

fishermen (*Fig 108*). They hold the live bait, taken from the main tanks by handnet, ready for use in the chumming operation. These small tanks, which in Japan are about 1m diameter and 70 to 100cm in height, were formerly made of wood but nowadays they are more commonly made of plastic (*Fig 109*). They are so designed that the water in them can easily be changed. This is generally done by means of a 25mm diameter perforated tube which extends around the wall at the bottom of the tank. The water spray system is used to pump water through the tube to be distributed in the tank through the perforations, an overflow outlet being fitted at the top of the tank.

7.4.3   BODY PROTECTION GEAR

During the angling operation, most pole-and-line fishermen wear a pad of heavy leather attached to a wide, adjustable belt to take the weight of the pole, while fishing. A U-shaped short section of 25 to 32mm diameter rope

Fig 106   Using small scoopnets for chumming on board a Japanese tuna vessel

is wired to the centre of the pad, the 'U' being wider than the butt of the pole.

The wearing of 'hard hats' (industrial-type protective helmets) is strongly recommended during angling.

Fig 107   Chumming with a small scoopnet in the Maldives

Fig 108 and Fig 109   Left, a deck chum tank and bucket, and right, a round chum tank on a Japanese vessel (foreground). Note the small scoopnet at the top of the tank

7.4.4 WATER SPLASHERS

In the smaller boats, where no mechanized water sprinklers are provided, the fishermen use all kinds of scoops to create the spraying and splashing effect to help with the attraction of tuna (*Fig 112*). The use of such scoops in Maldives is shown in *Fig 111*.

Fig 110   A scoop used in the Lakshadweep for splashing

7.4.5 ANGLING MACHINES

Automatic angling machines (*Figs 8, 9* and *112*) as used mainly on board large pole-and-line vessels, consist of a hydraulically operated FRP (fibre-glass) angling pole. The machine is able to perform three types of motion:

Fig 111   Photograph taken in the Maldives, just before fishing began. The two fishermen squatting either side of the rudder and holding scoops, ready to splash water, while the man sitting just aft of the mast holds a small scoopnet, with live bait

Fig 112   Automatic tuna angling machines. Left, American and right, Japanese

a jigging motion which keeps the hook moving to attract the fish; a rotating motion which brings the end of the pole inboard until the fish is on the deck, (this motion is triggered by the fish striking) and lastly, a jerking motion repeated until the fish is released from the hook.

Angling machines are installed on the vessel's rail and despite some minor drawbacks, are reported to be considerably more efficient than human anglers. Japanese made machines (*Fig 8*) which have their pole rotation axis at the rail are satisfactory for the Japanese type of vessels where there is ample space between the gunwale and superstructure, bait tanks, *etc.* Angling machines that have their pole rotation axis at the end of a 90 to 150cm long outboard-pointing boom are able to land the fish within the more confined deck space on board American-type tuna clippers, and these are manufactured in the USA (*Fig 112*).

The action of the automatic angling machines is briefly described above in 3.2.5. More detailed descriptions are provided in manufacturers' instruction manuals.

CHAPTER 8

THE FISHING OPERATION

8.1 Selection of fishing ground and search for tuna

The fishing operation starts, of course, with the search for schools of tuna. In the case of small-boat fishing in inshore waters the local fishermen usually know where and when to find the fish but in the case of offshore and distant water operations, deciding where to search may be a much more complex matter. Fishermen on offshore and ocean-going vessels study all available environmental data and fishing information before they determine the most promising ocean area.

8.1.1 ENVIRONMENTAL CONDITIONS

The hydrographic conditions favourable for fish concentrations in general, and for the creation of surface and near-surface tuna schools in particular, occur in areas where water masses of different temperatures meet, mix and form pronounced thermal fronts; where the thermocline (temperature layering) is comparatively shallow; where nutrient-rich waters are brought up from the depths by upwelling processes, and where local water turbulence develops around islands, peninsulas, over underwater banks, steep slopes and other topographic irregularities.

General information on oceanographic conditions, on the patterns of oceanic currents and on the seasonal fluctuations involved, is available in general oceanographic literature, atlases, *etc*. The influence of the environmental conditions on fish occurrence, abundance, migration and behaviour pattern have been widely dealt with in numerous reports of research institutes and in several books listed in the Bibliography.

Actual, current reports of conditions on the various fishing grounds and on the availability of fish can only be obtained from other fishing vessels by radio, and from research and scouting craft where such services are available. In any event, when a fishing vessel is on passage to distant fishing grounds, good radio contact with sources of such information is vital.

8.1.2 FISH BEHAVIOUR

Various tuna species in different areas may react differently to environmental conditions, each showing its peculiar behaviour pattern. Yellowfin, for example, often occur where the thermocline is shallow.

Migrating schools of skipjack approaching the coast or shallows, ascend to the surface water layer. They are generally dense and can be fished with relative ease. Hence, when experienced fishermen search for skipjack they tend to explore first the inshore waters around islands, reefs and banks, before sailing to the offshore areas and into the open ocean.

Bonitos, frigate mackerel and little tuna generally school more inshore than in offshore waters and are therefore accessible even to the smallest pole-and-line fishing boats.

8.1.3   Tuna spotting

When a fishing boat gets to the selected fishing ground, a watch is kept for signs of tuna schools. These signs include: flocks of birds which follow schools and prey on the small fish chased by the tuna (*Fig 113*); fish jumping out of the water; ripples on the water as though made by a breeze but actually caused by the fish swimming just below the surface, and feeding or 'boiling' schools where the tuna are ravenously feeding on schools of small fish (*Fig 114*).

Schools of tuna often follow floating objects—logs, wrecks, flotsam and jetsam and even large sea creatures such as big sharks and whales—swimming under and in the immediate vicinity of the object being followed.

On bigger vessels, an echo sounder watch is kept and information is continuously obtained by radio from other fishing vessels in the area, while environmental conditions are kept under study to find out if and when they

Fig 113   A school of tuna, accompanied by birds

Fig 114   A 'boiling' school of feeding tuna, with no birds visible

are favourable for concentrations of fish. Trolling lines, too, are set out.

There are variations from country to country in the way tuna are located and fished, these differences being due to local conditions and traditions. In Tahiti, for example, where live bait is not used, the fishermen generally locate schools by the presence and behaviour of sea birds. If the tuna are swimming deep down, the birds wait for them to come up, either circling around or resting on the water. The local fishermen say they can judge the size of the school by studying the behaviour of the birds and thus are able to prepare the necessary poles and the peculiar pearl-shell lures before the tuna come up.

8.2   The capture operation

8.2.1   APPROACHING A SCHOOL OF TUNA AND ATTRACTING THE FISH

When a school is found and the boat is close enough for the skipper to appraise the situation, he decides on the best approach. Such factors as the direction of the wind, the condition of the sea, the behaviour of the school and the direction in which it is moving as well as the position of the vessel all have to be taken into account. In general, it is considered that it is best for the boat to be leeward of the school.

The rate at which a school moves depends to a large extent on the species of small fish on which it is feeding. For example, if the tuna are feeding on whitebait they will slow down because whitebait are slow swimmers. But if they are feeding on sardines or mackerel they will speed up because these small fish are faster swimmers. If a school is congregating around floating objects, the normal method of approach is for the boat to circle around the school at slow speed, gradually getting nearer, then chumming with the live bait to attract the tuna.

If there is a flock of birds present, the boat is steered toward the head of the flock and then the crew chums with live bait to attract the fish to the vessel's side. Some believe that the important thing is to attract 'the leading fish' as this will help to keep the school around the boat. In the cases of both the submerged and jumping schools, the approach is similar. *Fig 115* illustrates schematically how a school of tuna should best be approached.

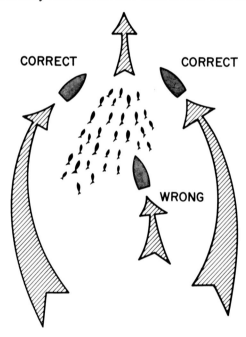

Fig 115   Correct and incorrect method of approaching a school of tuna

All preparations are made for the fishing operation before the chummer starts to throw the live bait out toward the school, seeking to form a thin but continuous chum line which will attract the leading fish toward the boat. As and when the fish come toward the vessel, chumming is increased in volume and speed.

If the tuna react well, the engine is stopped and water spraying is started. When some of the fish move from the stern to the side of the boat, large quantities of live bait are chummed to increase the attraction. When the fish are swimming round the boat they are considered fully attracted.

While a hungry school may react vigorously to the chumming and spraying, the fish can easily be scared away, for instance, by dropping a hooked fish back in the water or by allowing blood from the catch to fall into the sea.

8.2.1.1   *Spraying and splashing*

In the Maldives, Lakshadweep, Sri Lanka and other areas where small boats are used, the fishermen splash the water with all kinds of wooden spades to create spray when the tuna school is being chummed and fished (*Fig 110* and *111*). It is generally believed that the spray sounds and/or looks to the tuna as if caused by a school of small prey fish. Whether it is so

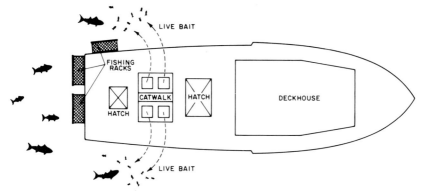

Fig 116   Chumming and maintaining contact with a school of tuna on a clipper type tuna boat

or not, it definitely helps to attract the fish and to keep them in the vicinity of the fishing vessel.

8.2.1.2   *Luring fish from deep water*

If a school of tuna occurs too deep to react to chumming with the usual surface swimming bait fish, it can be lured to the surface if a diving live bait, such as juvenile 50 to 70mm wrasse (*Labridae*) is available onboard. The wrasse tend to dive straight down when thrown in the sea. This is not a good behaviour for regular tuna chumming, but by throwing a single wrasse every few seconds the deep-swimming tuna is attracted by the diving fish and soon detect where they come from and ascend toward the surface. Regular live bait is then used for chumming, but initially it is a good practice to use hooks baited with live wrasse hooked by their tails and, while setting the hook, to simulate the diving behaviour of the fish.

8.2.1.3   *Flotsam*

An interesting way of attracting fish is sometimes used by the Tahitian fishermen who do not use live bait. If they find floating debris they get alongside and lift it aboard. The small fish which have been sheltering under it then get under the boat, which moves very slowly towards a school of tuna which attack the small fish in a sudden frenzy of feeding and are easily caught until, having devoured most of the prey and dispersed the rest, the remainder of the school go away.

The use of flotsam to attract fish has long been practised by Japanese pole-and-line fishermen also, and is being further developed in Hawaii. Specially constructed buoys and floating platforms are anchored offshore for the specific purpose of attracting fish. Such buoys are marked with radar reflectors and can be constructed of cheap, local materials. Good fish aggregations and improved catches have been reported.

8.2.2 ANGLING

The crew take up their angling stations as soon as a school is located, and the bait is put into the chumming tanks and chumming buckets. Depending on the type of vessel, the chummers take up their position at the bow, amidships and/or aft, while the fishermen with their poles and lines wait side by side on the sponson, racks or on the fishing deck (*Fig 109* and *111*). Where there are fishing machines, they are made ready for operation.

Fig 117 Angling for tuna in the Lakshadweep Islands, using pole-and-line with live bait. The two men sitting at each corner of the angling platform are splashing water to attract the fish. The correct position for the chummer is amidships and not as seen in the picture

Fig 118 Angling for tuna in Tahiti, using pearl-shell lures. No live bait is used in this type of operation

As already mentioned, the boat is manoeuvered so that the fishing side is on the lee of the school and when the boat is stopped and the water sprinkling system is started, fishing begins in earnest (*Figs 109, 117–120*).

The action which takes place from the moment the fish strikes to when it is safely on board, is of critical importance. However, the special skill involved cannot be learned from books, but only on the job and from accumulated experience. All that can be said here is that the tuna angling action is, unlike sport-type angling, an almost single-sweeping motion of jerk and lift, ending with another releasing jerk; the action is fast, strong and uncompromising, without any 'playing' or 'tiring' of the fish. A highly skilled fisherman can unhook the fish with a slight jerk of the line as it swings overhead and before it bumps on deck; if not, the release of tension on the line and the thudding of the fish on deck causes the hook to be freed. The operation may last 5 minutes, perhaps 10 but never more than about 20 minutes. It is a period of intense chumming and angling in an effort to keep the school around the boat for as long as possible and to catch as many tuna as possible before the school disperses.

Where the angling is done from the stern or mainly from the stern, the pattern of chumming and maintaining contact with the school may be as shown in *Fig 116*.

If the fish do not react favourably to the efforts of the fishermen to get them to bite, hooks are baited with live bait. This is done in one of three ways—by the gills, by the back or by the snout. Such small bait as anchovies

Fig 119   Fishing tuna in the Maldives. In the foreground, the chummer is throwing live bait; on the stern platform a man sits splashing water while the angler is ready to hook tuna from the school

are usually threaded by the gills while larger fish, such as sardines, are threaded by the snout. Care should be taken to thread the hook through the fish so that it will not slip off the hook easily yet will look lively and attractive to the tuna. The threading should be done quickly so as to avoid, as much as possible, damaging the bait. If the school is composed of big fish, one jig may be jointly used by two fishermen in the so-called two-pole fishing, their lines being connected to a joint snood (*Figs 95, 120* and *121*). Jigs are used when the fish react well but if the reaction of the tuna is poor, then baited hooks are used. Baited hooks can be used with both one and two-pole fishing.

8.2.3   OTHER PRACTICES

In the Maldive Islands the boats sail *through* the schools, the chummer throwing out live bait while others are fishing, some working the poles, others assisting in landing the fish and one helping the chummer by keeping him supplied. After the boats have sailed through the school, they turn quickly and sail back through. The boats complete a fishing trip in a day. According to one report the fishermen yell at the top of their voices during the fishing operation. The rather dubious explanation for this practice is that it 'helps to excite the fish and stimulate their appetite'.

In Tahiti, where no live bait is used, the fish are attracted by the skilled application of pearl-shell lures (*Figs 92* and *96*) and fast motor boats (*Figs 24, 25* and *102*) are employed. The whole capture operation is therefore markedly different.

Tahitian bonito boats are usually manned by two fishermen and a 'boy' or apprentice, although in some cases only two men operate the vessel. When a school is being worked, the 'boy' removes, clubs, bleeds and stacks fish in the racks while both fishermen work the poles and lines. One of the fishermen operates the tiller with his leg. The skipper of the vessel usually keeps a lookout for seabirds and interprets their tell-tale behaviour. All three crew help in general cleaning and other work on the boat when not fishing.

Schools of tuna are detected mainly by the accompanying flocks of birds. Following their prey the tuna dive and surface while the birds feed ravenously on the prey fish, soar, or sit on the water, waiting for the fish to reappear.

As soon as a school surfaces the fishermen take their boat to the edge of the school and slow down to one or two knots. If the tuna are biting well, they stop the boat. The two fishermen in the boat, standing in the stern and steadying themselves with one hand on the tiller, whip the water with their lures, dragging them from side to side and making them weave through the water as they draw them to the boat. When a fish strikes the fisherman hauls it clean into the boat with one steady movement. The 'assistant' removes the hook if necessary and the fishing immediately starts again. If

Fig 120 Pole-and-line fishing on a Japanese vessel. The anglers work in pairs to land large fish, using two-pole operation

the tuna being fished are more than 10kg they are too heavy to be pulled in by one man so two fishermen attach their lines to one lure and fish together.

If the fish do not respond to the lures, the fishermen quickly change to other poles and lines and try out various lures of differing colours and sizes until they find one that attracts the tuna. Short poles and small metal *champignon* jigs are used for fishing tuna of less than 2kg.

The testing of different lures has to be done very quickly, as the tuna schools seldom remain feeding at the surface for more than 15min. While there are theories about sizes and colours of lures in relation to the size of fish, most experienced fishermen think that 'only the fish can give the answer' so that the trial and error method is the most practical.

The fishermen plug the holes in the self-draining cockpit for the duration of the fishing operation. This is to prevent blood from the catch draining into the sea because, they say, the blood would cause the tuna to dive away. It would also attract sharks which could disperse the tuna. For the same reasons they never throw the entrails of the gutted fish overboard at the fishing ground.

8.2.4 TROLLING

Troll lines are used while searching for tuna, for when a fish strikes a trolling jig this may be an indication of the presence of a fishable concentration. The trolling operation may continue for some time, the vessel being steered on a circular course, while some live bait is scattered. If more fish appear, angling operations begin.

Trolling is also used as an auxiliary fishing method both on the way to and from the fishing grounds, and if angling operations render poor results. In some cases troll lines are used when the boat runs out of live bait, when

Fig 121   Two-pole angling of large tuna

the fish are plentiful and time, and space in the fish hold, permit fishing to continue. In this case a maximum number of lines should be set out, using outriggers. *Figs 122* and *123* give the specifications of gear and deck and outrigger arrangements as used in the South Pacific Ocean

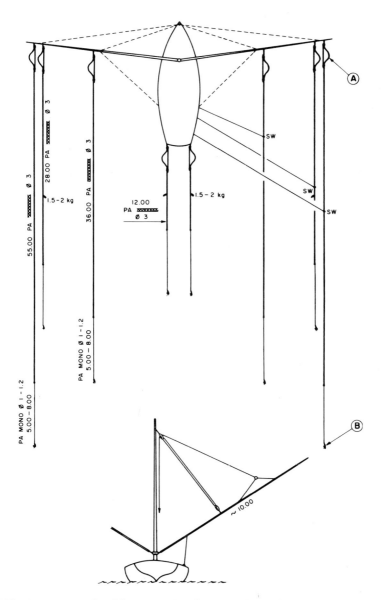

Fig 122   Arrangement of troll lines—a variant from New Caledonia

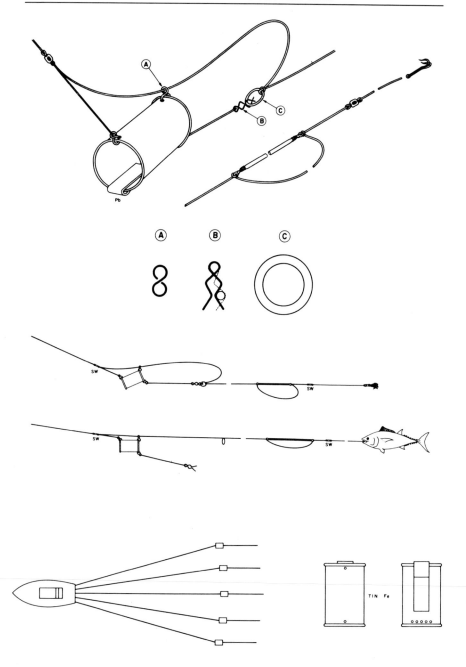

Fig 123   Arrangement for surface and deep trolling-variants from New Zealand showing a 'diving tin' which is released when a tuna takes the hook

CHAPTER 9

HANDLING FISH ON BOARD

Methods of handling the catch on board vary from doing little or nothing at all until it is landed in a port to the use of refrigeration for cooling or freezing it. Handling largely depends on the class of vessel and the type of fishery. Generally speaking, the small vessels operating in the inshore fishery may do no more than wash the fish, because it is landed and put on the market within a few hours of capture, whereas the distant water vessels have to hold the fish for many days, perhaps some weeks, so that they use sophisticated means to keep it from spoiling.

Whatever the method of handling the fish, care must be taken to avoid bruising them or causing skin abrasions.

9.1 Handling the catch in small-scale fishing

As previously mentioned, the handling of fish on board is often determined by local conditions. In Tahiti, for example, the fish are brought to harbour and sold the same day because such fresh fish fetch the best price. Any fish left unsold are usually stored in a chillroom and sold later at lower prices.

Handling on board consists of gutting and washing the catch as soon as possible, after which the carcasses are stacked, head down, in a removable rack in the shade (*Fig 124*). They are frequently doused with water to wash away the blood and keep them clean and looking fresh, but they are not iced. As detailed above (*8.2.3*) blood and entrails are not allowed overboard during fishing.

In the Maldives much of the fish, which is landed the day it is caught, is only washed on board, then cleaned, cooked and smoked ashore. This product is packed in bags for export. In Sri Lanka, too, the catch is landed the same day and is not usually processed in any way on board except to be washed, ice being rarely used at present. On the other hand, some of the skipjack fishing boats of Hawaii use ice to keep the catch cooled, the fish being stored in the emptied live-bait tanks. In Japan, the catch of the inshore vessels is consumed fresh—as raw fish—so that the trips are kept as short as feasible. The fish are handled with great care. They are washed and held, whole, in the holds (mainly emptied live-bait tanks) in a mixture of sea water and crushed ice. This slush not only cools the fish but acts as a cushion to prevent them from being crushed and bruised. This system keeps the tuna fresh for about four days, after which, if not landed, their quality deteriorates rapidly.

Since the catch of the smaller inshore boats is mostly sold daily to the fresh fish market many fishermen do not ice their fish. As the catch is usually landed only a few hours after capture, it does not present any difficulty in keeping it fresh. Where no proper fish holds or insulated ice boxes are available, the fish are kept in the shade, especially in tropical and subtropical areas, and are doused frequently with water to keep them clean and fresh.

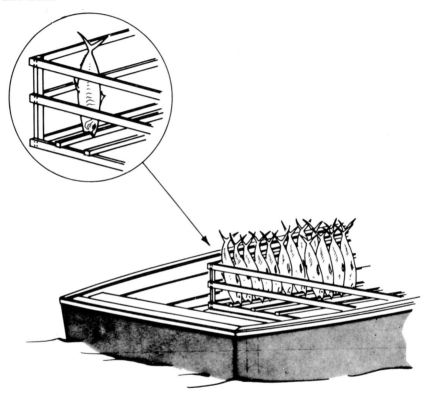

Fig 124   Tuna stacked on racks aboard a Tahitian *boutier* tuna boat

9.2   Fish handling on medium sized Japanese vessels

The vessels of this class also work mainly for the fresh (raw tuna) market. No ice is used but the fish are washed and transferred by conveyor belt or by hand to the four holds (each 2 to 2.5t capacity) where they are stored in chilled water which is changed every 24 hours. Brine, cooled by the freezer, is pumped through pipes in the holds to keep this water at the required temperature. In recent years a mixture of sea water and fresh water has been used in the bigger vessels for cooling and this has proved successful. Some of the medium size boats are now using this mixture.

9.3   Handling the catch on ocean-going Japanese tuna vessels

Most ocean-going vessels, equipped with refrigeration plants also have a conveyor belt to carry the catch to the freezing installation. The fish are first washed where they are deposited on deck when caught and again during transport on the belt. If there is no conveyor, then they are passed from hand to hand by the crew.

The fish are held in cold storage after one of two refrigeration processes, brine freezing or semi-airblast freezing. The big Japanese pole-and-line tuna fishing vessels have airblast freezers with a capacity of 5 to 10t a day. However, most of the freezing is done in the brine tank because it is a cheaper method for mass freezing. The outlay for an airblast installation is considerable because machinery such as high-powered freezers and auxiliary engine are costly.

Furthermore, a separate freezing room is needed. The practice, therefore, is to use the brine freezing method when a big catch of skipjack and albacore is made and the airblast freezer when the catch is small or consists of large tuna.

9.3.1   BRINE FREEZING

High-density brine is prepared by adding either calcium chloride or common salt to seawater and agitating it while the vessel is in port or on the way to the fishing grounds. The tanks are filled with brine and fish before being cooled down to $-18\,^{\circ}\mathrm{C}$ to $-20\,^{\circ}\mathrm{C}$ which takes 10 hours. If there is not enough fish to fill the tank after one fishing operation, the brine is cooled to only $0\,^{\circ}\mathrm{C}$ and to the full freezing temperature above only after more fish have been caught and the tank filled.

When the freezing of the fish in the brine tank is completed, the brine is pumped out. During the voyage the fish are kept in dry refrigeration, either in the same tank in which they were frozen or in another cold-store tank. One way to get the fish from the brine-freezing tank is to flood it with sea water. The frozen fish then float up and are taken out one by one.

Sometimes the fish stick to the bottom of the brine tank or to each other. They have then to be separated by hand very carefully as they are easily damaged. The cold holds are kept cool by the circulation of brine from a freezer tank through pipes installed in the holds, and the temperature is controlled by a thermostat. As it is easier to maintain the required temperature in a small rather than a large hold, a number of small holds are preferred to two or three big ones.

Also, the fish do not move so much in small holds, and this reduces the risk of abrasion and other damage. A discoloration of the flesh, known as 'orange meat', occurs sometimes in brine-frozen skipjack before or after canning and destroys the product's commercial value. Pre-cooling the fish before freezing is suggested as a means of preventing the discoloration but this seems to require a lot of time and extra work—more than 24 hours pre-cooling in a temperature of $2\,^{\circ}\mathrm{C}$ to $-6\,^{\circ}\mathrm{C}$. This corrective treatment, unfortunately, tends to reduce the freshness of the fish.

9.3.2 AIRBLAST FREEZING

In the semi-airblast freezing process, the bigger fish are first gutted, (the smaller ones are processed whole), and all are washed. They are then placed on galvanized iron sheet and put on the cooling grids in the freezer. The machine is closed and the fans switched on to circulate the air, the temperature in this process reaching −45°C to −60°C. The freezing is completed in 25 to 30 hours. The fish are then glazed and put in the cold storage holds.

9.4   Handling the catch on American tuna clippers

The large West Coast tuna clippers were mostly phased out many years ago and have been replaced by purse seiners. However the method of catch handling which was evolved while they were still in operation is still popular on purse seiners and on the remaining—and a few new—tuna clippers or 'bait boats'. This can be attributed to the method's extreme simplicity and, in spite of criticism on the part of food technologists, because the quality of the product landed is good enough for canning.

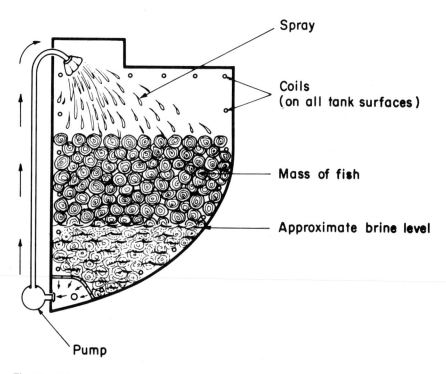

Fig 125   Brine spray system as used on American tuna clippers

Fish are frozen in wells using the brine spray method by which chilled, saturated brine is circulated from the bottom of the well and sprayed over the mass of the fish (*Fig 125*). No pre-chilling is employed, and the catch is loaded directly into the freezing wells.

Usually a sack is tied over the brine discharge, which is in the hatch coaming, to disperse the pump flow, or considerable foaming occurs.

Cold brine runs over the surface of the exposed fish and then down through the mass. Much of the brine goes to the sides of the well and runs down over the coils which are attached to the sides. The fish are held away by the coils and therefore this portion of the brine is chilled. The heat transfer rate from the coils is low because of by-passing and low fluid velocity, but it is sufficient eventually to freeze the fish.

Full tanks of fish are circulated until the last fish added are frozen, then the brine is pumped out into another tank and the frozen fish are held dry using the coils to maintain freezing temperature. Frozen tuna are held at $-7\,^{\circ}$C ($20\,^{\circ}$F) or lower in the wells.

By this method (*Fig 125*) the fish at the bottom of the well are totally submerged in the brine for a prolonged period and thus subjected to salt penetration more than those in the top layers. This non-uniformity of treatment and hence, of the resulting product, may be one of the reasons for the criticism. However, if better drainage was provided, and efficient brine cooling to assure the desired $-18\,^{\circ}$C to $-21\,^{\circ}$C temperature of the brine, and if good air circulation could be provided for the post-freezing cold storage stage, this system should satisfy most requirements.

The refrigeration capacity of brine systems for small tuna vessels must be matched to the maximum catch rate, since tuna spoil rapidly in the ambient conditions.

External heat exchangers, if used, must be carefully designed to avoid premature ice formation and to provide high efficiency of heat transfer. Metals containing copper *must not* be used in refrigeration equipment for chilling or freezing tuna, since canned tuna meat is discoloured by even slight traces of copper. Galvanized steel heat exchangers are suitable but have a limited service life, so that the use of stainless steel is highly recommended.

For small boats and storage periods of one or two days tank coils are not used if external heat exchangers are installed, but the tanks must be very well insulated. The heat exchangers should be mounted in a reasonably protected area and preferably *not* in the engine room.

The various brine freezing methods are recommended only for tuna-like fish. Other species may suffer from excessive salt penetration into the flesh. Even tuna can be thus affected if the brine is not cold enough, and minimum temperatures of $-18\,^{\circ}$C to $-21\,^{\circ}$C are therefore recommended.

WORLD GUIDE TO BAIT FISH FOR POLE AND LINE FISHING

Fishing areas—see column 3

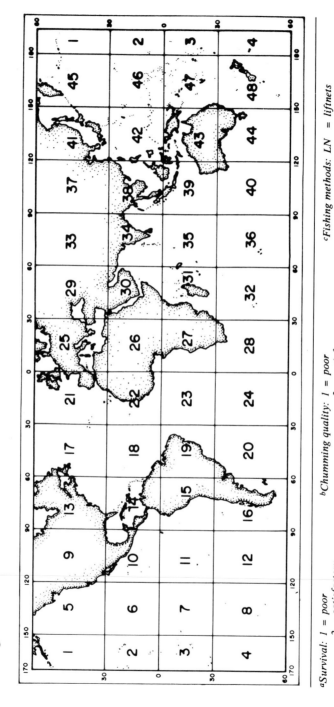

aSurvival: 1 = poor
2 = satisfactory
3 = good
4 = excellent

bChumming quality: 1 = poor
2 = satisfactory
3 = very good
4 = excellent

cFishing methods: LN = liftnets
DN = drive-in nets
SN = surrounding
nets
LT = light attraction
BS = beach seines

Scientific name	English name	Area fished	Survival[a]	Chumming quality[b]	Fishing methods[c]	Remarks
Apogonidae	Cardinal fishes	42			DN	
Apogon semilineatus	Striped cardinal fish	41 42 38 34 39	3	2	DN	Only in small quantities
Apogon thermalis		34 35		1	DN	
Archamia lineolata		30 34 38 31 35 39 41 42	3	2	LN	Only in small quantities
Rhobdamia cypselurus	Cardinal fish	2 46	3	4	DN	
Atherinidae	Silversides					
Allaneta forskali	Hardyhead	34 38 39 43	3	1	BS LT	Considered live-bait fishes; easily kept in captivity for extended periods
Allaneta ovalaua	Fijan silverside	47 3	2	2		
Allaneta valenciennei		42 46	1	1		
Allaneta woodwardi	Silverside (teber)	42		1		
Atherina afra	Silverside	31				
Atherina insularum	Iao silverside	41 2	2	4		
Atherina presbyter	Silverside	21 22				
Pranesus duodecimalis		34	3	3		
Pranesus insularum	Hawaiian silverside	2	4	2		
Pranesus pinguis	Silverside	43 47 3 46 45 1 39 35 2 34 31 30	3			
Albulidae	Bonefishes					
Albula vulpes	Bonefish	2	2	4	BS	Juveniles only; school along sandy shores and on reef flats
Caesiodidae	Fusiliers					
Caesio chrysozonus	Goldband fusilier	42 41 1 2 3 47 43 39 35 34 31 30	3	4	LT	Juveniles only; occur in small quantities
Caesio coerulaureus		42 41 34	3	4		
Caesio erythrogaster		34				

WORLD GUIDE TO BAIT FISH FOR POLE AND LINE FISHING—continued

Scientific name	English name	Area fished	Survival[a]	Chumming quality[b]	Fishing methods[c]	Remarks
Caesio tile	Fusilier	41 42	2		LT	Juveniles only, occur in small quantities
Gymnocaesio argenteus	Snapper	2 46 47	3	4		
Gymnocaesio gymnopterus		43	3	3		
Pterocaesio pisang		43	3			
Carangidae	Jacks, horse mackerels					
Carangoides malabaricus	Malabar cavalla	31 30 34 38 39 42 47 46 41			LT DN	Juveniles only
Chorinemus tol	Leatherskin	2 47 3				
Decapterus dayi	Scad	31				
Decapterus maruadsi	Round scad	41 42 38 43 39 34 30 31			LT	Juveniles only
Decapterus pinnulatus	Mackerel scad	2				Juveniles only
Decapterus sp.	Scad, round scad	10 14				Juveniles only
Selar crumenophthalmus	Bigeye scad	2 3 47 46 42 30 31 34 38	3	1	LT	Juveniles only
Selaroides leptolepis	Yellowstripe trevally	39 42 46 47				Juveniles only
Trachurops sp.	Scad mackerel	3 47			LT	Juveniles only
Trachurus japonicus	(Pacific) Horse mackerel	41 45 42 46 47 48 1 2 3 43 39 35 31 30 34			LT	Juveniles only
Trachurus trachurus	(Atlantic) Horse mackerel	21 22 25 29			LT	Juveniles only
Chanidae	Milkfishes					
Chanos chanos	Milkfish	2 3 46 47 42 43 38 39 30 34	4	3	BS	Juveniles only
Cichlidae	St. Peter's fishes Cichlids					
Tilapia mossambica	Tilapia	Cosmopolitan	4	3		Juveniles only, experimentally cultured

	Common name				Experimental	Seasonality of abundance
Clupeidae	Herrings					
Dorosoma petenense	Threadfin shad	2		4		
Dussumieria acuta	Rainbow sardine	1 2 3 46 / 47 42 43 / 38 39 34 / 30 31	1	3		
Etrumeus micropus	Pacific round herring	2				
Etrumeus teres	Round herring	41 38 42 47 / 48 44 40 39 / 5 9 10	1		LT	
Harengula sp.	Sprat, sprat herring	10 14				
Herklotsichthys punctatus	One-spot herring	43 42 3 47 / 39 35 31 28 / 30 34 38	1		LT	
Jenkinsia lamprotaenia	Dwarf herring	14				
Opisthonema libertate	Pacific thread herring	10 9 14	2			
Sardinella leiogaster	Smoothbelly sardinella	43 42 46 39 / 38 34 31 30	1		LT	Preferably small
Sardinella sirm	Spotted sardinella	31 30 34 38 / 39 43 42 46 / 47	4	3	LT	
Sardinops sagax caerulea	California sardine	9 10 5				
Sardinops melanosticta	Japanese sardine	45 41 42			SN	
Sardinops neopilchardus	Galapagos sardine	47 48 44 40 / 4 11	3	3		
Spratelloides delicatulus	Delicate round herring	3 47 2 46 42	1	3	SN LT DN	Easy to catch because do not dive when surrounded
Spratelloides gracilis		43 39 38 34 / 35 31 32 28 / 47 43 42 46 / 38 39 34 30 / 31 41			LT	
Emmelichthyidae						
Dipterygonotus	Redbait					
leucogrammicus	Redbait	34			LN	Over rocks 9–18m deep

WORLD GUIDE TO BAIT FISH FOR POLE AND LINE FISHING—*continued*

Scientific name	English name	Area fished	Sur-vival[a]	Chumming quality[b]	Fishing methods[c]	Remarks
Elopidae	Tarpons					
Elops machnata	Tarpon	34 35		2		Juveniles only
Engraulidae	Anchovies					
Anchoviella guineensis	Guinean anchovy	22 26 27				
Cetengraulis mysticetus	Anchoveta	5 9 10 14 15	3	4	LN	Hardy, withstands crowding
Engraulis australis	Australian anchovy		3	3		Extremely prone to panic
Engraulis japonicus	Japanese anchovy	45 41 38 42	4	4	LT LN SN	Resistant to both oxygen deficiency and death from confinement
Engraulis mordax	Northern anchovy	5 9 10 14	3	4		Not as hardy as anchoveta
Engraulis ringens	Southern anchovy	14 10 15				
Scutengraulis mystax (or *Thryssa mystax?*)		38 42	3	4		
Stolephorus bataviensis	Batavia anchovy	3 47 46 42 / 38 39 34		4		
Stolephorus buccaneeri	Buccaneer anchovy	3 47 46 42 / 38 39 34 30 / 31 32 28 41 2	1	3	LT	Excellent bait fish with both good attractability and handling qualities but unfortunately its occurrence is unpredictable
Stolephorus devisi	DeVis's anchovy	47 43	2	4	LT	Excellent bait fish but delicate and difficult to transport
Stolephorus heterolobus	Shorthead anchovy	42 43 47 46 / 3 38 39 34 / 30 31	1	3	LT	Delicate, requires extreme care in handling
Stolephorus indicus	Indian anchovy	47 43 3 46 / 42 38 39 34 / 30 31	1	4		Delicate
Stolephorus purpureus	Nehu	2	1-2	2-3	SN LT	Considered delicate with mortalities following capture up to 25 percent per day
Thrissina baelama	Baelama anchovy	43 47 3 39 / 38 34 35 30 / 31	1-2	1	BS LT	Hardy but loses its scales easily and must be handled with care

Family / Species	Common name					Remarks
Kuhlidae *Kuhlia sandvicensis*	Mountain basses Mountain bass	2				Juveniles make good live bait
Mugilidae	Grey mullets	Cosmopolitan many species	2–3	4	LT BS	Juveniles only, availability and suitability vary with species and area
Mullidae	Red mullets Goatfishes	Cosmopolitan many species	2	1–3		Conflicting reports as to their chumming quality
Poeciliidae *Mollienesia latipinna* *Poecilia mexicana*	Livebearers Sailfin, molly Molly, topminnow	2 3	3	1	BS	Used only when other bait unavailable To be cultured
Polynemidae *Polydactylus sexfilis*	Threadfins Threadfin	2	2–3	3	BS	Only small ones suitable
Pomacentridae	Damselfishes					Secondary bait fish, used when more desirable species are scarce
Chromis caeruleus	Puller	34	3			
Lepidozygus tapeinosoma	Coral fish	34				
Pomacentrus pavo	Damselfish	46 3 2 42				
Daya jerdoni	Damselfish, green puller	34				
Scombridae *Rastrelliger kanagurta*	Mackerels Indian mackerel	42	3	1	LT	Juveniles only, in tanks prey on smaller fish
Scomber japonicus	Chub mackerel	41	3		LT	Juveniles only
Scomber japonicus tapeinocephalus	Spotted mackerel	42 41				
Scomber scomber	Mackerel	21 22	3		LT SN	Juveniles only

148

REFERENCES

ANON, *Tuna fishing with pole and line using live-bait (A pesca dos atuns i linha con isco
1950 vivo). Bol. Pesca, Lisbon,,* 7(28): 65–70.
ANON, *Live-bait equipment on Pacific tuna clippers. Nat Fisherman,* Nov issue: 11–2, 38
1954
ANON, *New plywood bait boat 'Aquarius' features novel method of loading fish. Nat
1956 Fisherman,* 37(10): 13
ANON, *Japanese robot pole for tuna fishing. Fishing News Int,* 9(10): 73–4
1970
ANON, *Seattle again in 1977 for US Fish Expo. Fishing News Int,* 14(12): 25–6, 28
1975
BALASUBRAMANYAN, R, *Studies on fish baits, Part 1. A note on the use of different baits for
1964 sea-fishing. IPFC Occas Pap,* (64/7): 8 p.
BEN-YAMI, M, *Fishing with Light. FAO Fishing Manuals,* Fishing News Books Ltd, Farnham,
1976 England
BEGNIS, R, *et al., Poissons de Polynésie. Papeete, Les Editions du Pacifique,* p 368
1973
CARLSON, C B, *Fishing vessel live-bait equipment. Commercial Fisheries Review,* 16(1): 7
1954
DELAPORTE, F, *Les populations de germons Thunnus alalunga dans le Nord-Est Atlantique.
1973 Thèse de Doctorat d'Etat et Sciences Naturelles.* Institut Scientifique et Technique des
Pêches Maritimes, Nantes, pp. 99–194
DEVAMBEZ, L C, Personal communication
1976
DOUTRE, M, *La Pêche du thon a l'appât vivant en Californie. Rev Trav Inst Pêches Marit,
1956 Nantes,* 20(4): 449–74
DUPONT, E and RALISON, A, *Etude de la pêche à la bonite à l'appât vivant à Madagascar.
1973 Doc Tech Proj Dev Pêches MDR/PNUD/FAO,* (MAG/68/515 No 9)
FAO, *Local tuna fishery, Suva, Fiji. Livebait pole-and-line fishing for tuna.* Report prepared
1974 for the Government of Fiji by FAO of the UN acting as executing agency for the
UNDP. Based on the work of R E K D Lee, Rome, FAO/UNDP, FI: DP/FIJ/70/504,
Tech Rep (1)
FAR SEAS FISHERIES RESEARCH LABORATORY, *Report on fishery survey in Bismarck and
1969 Solomon Seas by the research vessel 'Shunyo-Maru',* Oct—Dec 1968 *S Ser Far Seas
Res Lab,* (1): 170p (in Japanese, with English summary)
GOPALAKRISHNAN, V, *Status and problems of culture of baitfish for the skipjack fishery in
1976 the Pacific Region.* Paper presented to the FAO Technical Conference on Aquaculture,
Kyoto, Japan, 1976. Rome, FAO, FIR: AQ/Conf./76/E2
GUSTAFFSON, N, Personal communication.
1966
HESTER, F J, *Some considerations of the problems associated with the use of live bait for
1974 catching tunas in the tropical Pacific Ocean. Mar Fish Rev,* 36(5): 1–11
HESTER, F J and OTSU, T, *A review of the literature on the development of skipjack tuna
1973 fisheries in the central and western Pacific Ocean. NOAA Tech Rep NMFS (Spec Sci
Rep Fish Ser),* (661): 13 p
IKEHARA, I I, *Live-bait fishing for tuna in the Central Pacific. Spec Sci Rep US Fish Wildl
1953 Serv,* (107): 20 p
IVERSEN, R T B, *Use of threadfin shad, Dorosoma peteneuse, as live bait during experimental
1971 pole-and-line fishing for skipjack tuna, Katsuwonus pelamis, in Hawaii. NOAA Tech
Rep NMFS (Spec Sci Rep Fish Ser),* (641): 10 p
IWASHITA, T T, *Development and design of the Hawaiian fishing sampans.* Paper presented at
1956 the meeting of the Hawaii Section of the Society of Naval Architects and Marine
Engineers, January 1956, 27 p (mimeo)

JOHNSTON, H and O'GRADY, A, *How to rig tuna fishing gear. Fisheries Newsletter, Aust*, 1951   10(9): 7

JONES, S and SILAS, E G, *Synopsis of biological data on skipjack Katsuwonus pelamis* 1963   *(Linnaeus) 1758 (Indian Ocean). FAO Fish Rep*, (6)vol. 2:663–94

JONES, W, *The tuna live-bait fishery of Minicoy Island. Indian J Fish*, 5(2): 300–7 1958

JUNE, F C, *Preliminary fisheries survey of the Hawaiian-Line Islands area. Pt 3. The live-bait* 1951   *skipjack fishery of the Hawaiian Islands. Commer Fish Rev*, 13(2): 1-18

JUNE, F C and REINTJES, J W, Common tuna-bait fishes of the central Pacific. *Res Rep US* 1953   *Fish Wildl Ser* , (34): 54 p

KEARNEY, R E, LEWIS, A D, and SMITH, B R, *Cruise report Tagula 71–1. Survey of skipjack* 1972   *tuna and bait resources in Papua New Guinea waters. Res Bull Dep Agric Stock Fish, Papua New Guinea*, (8): 145 p

KOYAMA, T, *Fishing jigs in Japan with special reference to an artificial bait made of latex* 1959   *sponge rubber.* Modern Fishing Gear of the World, edited by H Kristjonsson. Fishing News (Books) Ltd, Farnham, England, pp 567–70

LE DRÉZEN, L, Personal communication 1976

LEE, R E K D, *The Philippines. Live-bait pole-and-line fishing for tuna.* Rome, FAO, FI: 1977   PHI/72/002/1: 26 p

LE GUEN, J C, POINSARD, F and TROADEC, J P, *The yellowfin tuna fishery in the eastern* 1965   *Tropical Atlantic (Preliminary study). Commer Fish Rev*, 27(8): 7–18

LEWIS, A D, SMITH, B R, and KEARNEY, R E, *Studies on tunas and baitfishes in Papua New* 1974   *Guinea waters. 2. Res Bull Dep Agric Stock Fish, Papua New Guinea*, (11): 112 p

LORIMER, P D, *Live bait tuna gear for smaller vessels. Fish Newsl, Aust*, 18(7): 13, 15 1959

MURAMATSU, S, *Pole-and-line fishing: deck design and equipment.* Fishing Boats of the World, 1960   edited by J O Traung. Fishing News (Books) Ltd, Farnham, England, vol. 2: 84–93

NAKAMURA, H, *Tuna: distribution and migration.* Fishing News (Books) Ltd, Farnham, 1969   England, 84 p

NORDHOFF, C, *Notes on the off-shore fishing of the Society Islands. J Polynes Soc, Wellington*, 1928   39(2–3)

POWEL, R, *The Hawaiian opelu net: successful introduction into the Cook Islands.* Rarotonga, 1961   Government of the Cook Islands, Fisheries Division, 6 p (mimeo)

REED, W, *Tahitian pole-and-line fishing for skipjack using pearl-shell lures.* 23 p (MS) 1975

SHANG, Y C, and IVERSEN, R T B, *The production of threadfin shad as live bait for Hawaii's* 1971   *skipjack tuna fishery: an economic feasibility study.* Honolulu, University of Hawaii, Economic Research Center, 42 p.

SHOMURA, R S (ed), *Collection of tuna bait fish papers. NOAA Tech Rep NMFS Circ*, (408): 1977   8–31

UCHIDA, R N, *The skipjack tuna fishery in Palau. The Kuroshio; a symposium on the Japan* 1970   *Current*, edited by J C Marr. Honolulu. East-West Center Press, pp. 569–82

—— *Studies on skipjack in the Pacific. Recent development in fisheries for skipjack tuna,* 1975   *Katsuwonus pelamis, in the Central and Western Pacific and the Indian Oceans. FAO FishTech Pap* , (144): 1–57

WALDRON, K D, *Synopsis of biological data on skipjack Katsuwonus pelamis (Linnaeus)* 1963   *1758 (Pacific Ocean). FAO Fish Rep*, (6)vol 2: 695–748

WILSON, P, *Live-bait skipjack fishing. Offshore commercial fishing in Micronesia.* Saipan, 1962   Office of the Director of Agriculture and Fisheries, Trust Territory of the Pacific Islands

YOSHIDA, H O, *Tuna fishing vessels, gear and techniques in the Pacific Ocean.* Proceedings of 1966   the Governor's Conference on Central Pacific Fishery Resources, Honolulu-Hilo, February 1966, edited T A Manar. Honolulu, US Bureau of Commercial Fisheries, pp 67–89

150

Other books published by Fishing News Books Ltd Farnham, Surrey, England.

Free catalogue available on request

Advances in aquaculture
Advances in fish science and technology
Aquaculture practices in Taiwan
Atlantic salmon—its future
Better angling with simple science
British freshwater fishes
Commercial fishing methods
Control of fish quality
Culture of bivalve molluscs
Echo sounding and sonar for fishing
The edible crab and its fishery in
   British waters
Eel capture, culture, processing
   and marketing
Eel culture
European inland water fish:
   a multilingual catalogue
FAO catalogue of fishing gear designs
FAO catalogue of small scale
   fishing gear
FAO investigates ferro-cement
   fishing craft
Farming the edge of the sea
Fish and shellfish farming
   in coastal waters
Fish catching methods of the world
Fish inspection and quality control
Fisheries of Australia
Fisheries oceanography
Fishermen's handbook
Fishery products
Fishing boats and their equipment
Fishing boats of the world 1
Fishing boats of the world 2
Fishing boats of the world 3
The fishing cadet's handbook
Fishing ports and markets
Fishing with electricity
Fishing with light

Freezing and irradiation of fish
Handbook of trout and salmon diseases
Handy medical guide for seafarers
How to make and set nets
Inshore fishing: its skills, risks, rewards
The lemon sole
A living from lobsters
Marine pollution and sea life
The marketing of shellfish
Mending of fishing nets
Modern deep sea trawling gear
Modern fishing gear of the world 1
Modern fishing gear of the world 2
Modern fishing gear of the world 3
More Scottish fishing craft
   and their work
Multilingual dictionary of fish
   and fish products
Navigation primer for fishermen
Netting materials for fishing gear
Pair trawling and pair seining
   –the technology of two boat fishing
Pelagic and semi-pelagic trawling gear
Planning of aquaculture development
   –an introductory guide
Power transmission and automation for
   ships and submersibles
Refrigeration on fishing vessels
Salmon and trout farming in Norway
Salmon fisheries of Scotland
Seafood fishing for amateur and
   professional
Stability and trim of fishing vessels
The stern trawler
Textbook of fish culture: breeding and
   cultivation of fish
Training fishermen at sea
Trout farming manual
Tuna: distribution and migration